URBAN POLICY EVALUATION:
CHALLENGE AND CHANGE

URBAN POLICY EVALUATION: CHALLENGE AND CHANGE

edited by
Robin Hambleton and Huw Thomas
University of Wales College of Cardiff

P·C·P
Paul Chapman
Publishing Ltd

Selection and editorial material copyright © 1995, Robin Hambleton and Huw Thomas. All
other material © as credited.

Paul Chapman Publishing Ltd
144 Liverpool Road
London
N1 1LA

British Library Cataloguing in Publication Data

Urban Policy Evaluation: Challenge and Change
   I. Hambleton, Robin II. Thomas, Huw
   307.1416

ISBN 1–85396–271–6

Typeset by Palimpsest Book Production Limited, Polmont, Stirlingshire
Printed and bound by Athenaeum Press, Gateshead, Tyne & Wear

A B C D E F G H   9 8 7 6 5

CONTENTS

ACKNOWLEDGEMENTS

This book comprises revised versions of papers given at two two-day seminars on urban policy evaluation supported by the Economic and Social Research Council (ESRC) and held in Cardiff in September 1993 and September 1994. Without the financial support of the ESRC it would have been impossible to run these seminars and without the seminars, funded by ESRC grant A 451 26 4003 24, there would be no book. Our first thanks go, therefore, to the ESRC for its support and encouragement.

We would like to thank all the participants in the two seminars, not just those who gave papers, but also colleagues from local and central government and elsewhere who attended and offered critical comments and insights. The opportunities for an in-depth dialogue between practitioners and academics as well as between academics with different disciplinary perspectives are comparatively far and few between. We are most grateful to everyone involved for engaging in a fair and frank exchange about alternative approaches to urban policy evaluation. In relation to the seminars we would like to offer a special thank you to Dee Gilmore, the Manager of the Short Course Unit in our Department, who organised the seminars with her usual skill, flair and enthusiasm.

Apart from our own contributions, some twenty authors have written chapters for this book, several of them working in pairs. We are delighted that everyone we approached agreed to write a chapter and we are most grateful for their efforts. We hope that they feel that the product is worthwhile.

Last but by no means least we want to thank our support staff on this project. Andrew Edwards solved computer hitches and Janice Cole helped with a number of the illustrations. Our secretary for this project, Jane Melvin, exhibited a cool mastery of the keyboard (imagine working for twenty-two academics all with their own foibles!), disarming patience with the editors and a good sense of humour throughout. We could not ask for more.

Robin Hambleton
Huw Thomas
Department of City and Regional Planning
University of Wales, Cardiff

November, 1994

LIST OF CONTRIBUTORS

Martin Boddy, Professor of Urban and Regional Studies and Director of the School for Advanced Urban Studies, University of Bristol

Philip A Booth, Senior Lecturer, Department of Town and Regional Planning, University of Sheffield

Michael Bradford, Senior Lecturer in Geography and Dean for Undergraduate Studies, Arts Faculty, University of Manchester

Paul Burton, Research Fellow, School for Advanced Urban Studies, University of Bristol

Michael Chapman, Lecturer, School of Planning and Housing, Edinburgh College of Art/Heriot-Watt University

Simin Davoudi, Research Associate, Centre for Research in European Urban Environments, Department of Town and Country Planning, University of Newcastle

Howard Green, Professor of Urban Planning and Head of Research Development, Leeds Metropolitan University

Robin Hambleton, Professor of City and Regional Planning, Department of City and Regional Planning, University of Wales, Cardiff

Annette Hastings, Research Fellow, Centre for Housing Research and Urban Studies, University of Glasgow

Patsy Healey, Professor and Director of the Centre for Research in European Urban Environments, Department of Town and Country Planning, University of Newcastle

Rob Imrie, Lecturer, Department of Geography, Royal Holloway University of London

Christine Lambert, Senior Lecturer, Faculty of the Built Environment, University of the West of England, Bristol

Steve Martin, Principal Research Fellow, Local Government Centre, Warwick University

Andrew McArthur, Lecturer in Urban Studies, Department of Social and Economic Research, University of Glasgow

Peter B. Meyer, Professor of Urban Policy and Economics, Centre for Environmental Management, College of Business and Public Administration, University of Louisville

Kevin Morgan, Professor of European Regional Development, Department of City and Regional Planning, University of Wales, Cardiff

Nick Oatley, Senior Lecturer, Faculty of the Built Environment, University of the West of England, Bristol

Graham Pearce, Lecturer, Public Sector Management Group, Aston Business School, Aston University

Brian Robson, Professor of Geography and Pro-Vice-Chancellor, University of Manchester

Murray Stewart, Professor of Urban Government, School for Advanced Urban Studies, University of Bristol

Huw Thomas, Lecturer, Department of City and Regional Planning, University of Wales, Cardiff

Cecilia Wong, Lecturer, Department of Planning and Landscape, University of Manchester

1

URBAN POLICY EVALUATION –
THE CONTOURS OF THE DEBATE

Robin Hambleton and Huw Thomas

As conditions in the inner areas become worse, symptoms of deprivation, such as crime and vandalism, assume exaggerated importance and remedies also shift from normal municipal programmes to the imposition of law and order.

(McConaghy, 1971, p. 356)

During the weekend of 10–12 April the British people watched with horror and incredulity an instant audio-visual presentation on their television sets of scenes of violence and disorder in their capital city, . . . a few hundred young people attacked the police on the streets with stones, bricks, iron bars and petrol bombs, demonstrating to millions of their fellow citizens the fragile basis of the Queen's peace.

(Scarman, 1981, p. 13)

The violent eruption of one of Britain's most deprived estates forced law and order onto the pre-election political agenda yesterday, . . . The Government and opposition condemned the five hour rampage of looting and arson at Meadow Well, North Tyneside . . . but there were calls from church leaders, residents and some politicians to recognise the hopelessness and despair.

(Wainwright, headline article in *The Guardian*, 11 September 1991)

These three quotes – from 1971, 1981 and 1991 – highlight not just the enduring existence of urban deprivation in British cities but also the continuing potential for social tensions to flare up into outbreaks of violence on the streets. The first quote, which sounds as if it appeared yesterday, was actually written over twenty years ago by Des McConaghy when he was working in the inner area of Liverpool as leader of the Shelter Neighbourhood Action Project (SNAP). He documented the miserable living conditions in Granby and correctly identified the potential for urban unrest in Liverpool (McConaghy, 1971).

The second quote, from ten years later, provides a reminder of the horrific violence which erupted in Brixton in 1981. And it was, of course, not just Brixton. The previous year had seen major disturbances in the St Paul's area of Bristol and in July 1981 there were riots in Southall, followed in rapid succession by major outbreaks of violence at Toxteth in Liverpool, Moss Side in Manchester and Handsworth in Birmingham – as well as a string of smaller-scale disorders in Southampton, Leicester, Nottingham, Derby, Wolverhampton, Bradford, Halifax, Leeds, Huddersfield, Blackburn, Preston, Teesside and elsewhere, including Hackney (Harrison, 1983). Lord Scarman identified two sets of explanations for the disorders in Brixton (Scarman, 1981). First, the riots could be viewed as a reaction to oppressive policing over a period of years. Second, the troubles could be seen as a protest against society by people, deeply frustrated and deprived, who sought to draw public attention to their grievances.

In 1991, a decade after Michael Heseltine announced his private sector orientated inner city policy to the House of Commons, the country was faced with a further wave of outbreaks of public disorder not just in inner city areas – for example, the Meadowell estate in North Tyneside and Newcastle's West End – but also in peripheral housing estates, such as Ely in West Cardiff and Blackbird Leys on the edge of Oxford. The differences between these confrontations should not be ignored. A recent analysis of these disturbances has shown that the incidents which sparked off the riots were varied (Campbell, 1993). But there are common themes: all took place in the poorer parts of cities, all involved young people, many of them unemployed, and all would seem to suggest that something is seriously wrong with existing policies for cities.

In the years since 1968 central governments of diverse political persuasions have pursued a variety of urban initiatives. If outbreaks of social unrest can be taken as a performance indicator, and it is difficult to see how rioting in the streets can be viewed as an irrelevant signal in the context of urban policy evaluation, then the evidence suggests that these many urban initiatives have been less than successful. However, this is only one indicator of urban policy performance. Indeed some commentators would argue that urban policy is less to do with the amelioration of social ills than with the promotion of economic regeneration in areas which have found it difficult to attract inward private investment. Certainly in the years since 1979 the government has emphasised private sector property development as an approach to urban regeneration believing that such a strategy would stimulate wider economic benefits.

Thus, vast sums of public money have been channelled into schemes designed to pave the way for private sector projects, particularly in the areas managed by the urban development corporations (Imrie and Thomas, 1993). The slump in the property market in the early 1990s showed how this strategy was somewhat unbalanced. The spectacular financial collapse, in May 1992, of Olympia and York, owners of the colossal Canary Wharf development in London Docklands, highlighted vividly the fragility of the property-based approach to urban regeneration. Thus, in 1992 the *Financial Times* noted that London docklands was not only viewed as a planning and architectural mess, but also as 'the most costly disaster in the history of UK property development, following a collapse in property values and a glut of vacant property' (Houlder, 1992).

The limitations of an approach which put too much emphasis on private sector property development came to be recognised and partly explains recent shifts in national urban policy. The *City Challenge* initiative, launched in 1991, the Single Regeneration Budget (SRB) and the development of Integrated Regional

Offices (IROs) in the English regions, introduced in 1994, suggest that the government may be attempting to broaden the balance and thrust of urban policy. Certainly there appears to be a growing recognition of the part that local authorities can play and of the potential for developing collaborative working between the public, private and voluntary sectors.

These opening remarks introduce three themes which are developed below and explored in various ways in subsequent chapters. First, urban policy is an important part of public policy. The nature of the problems now confronting cities in the UK suggests that a many sided and versatile urban policy response is required – one that taps the energies of local communities, the local authorities and the private sector as well as involving action at national level. Second, the aims and objectives of, even the whole approach to, urban policy are contested. Different interests define the urban policy agenda in different ways to the point where there may be total disagreement. For example, the differences of view between central government and urban authorities in the 1980s led Whitehall to develop a variety of measures designed to exclude the elected local authorities from the urban policy process. Third, developments in urban policy – or at least the public announcements about developments in urban policy – have been frequent. The history of urban policy since 1968 is littered with examples of new initiatives. This proliferation of urban initiatives has been criticised in numerous studies including an assessment by the Audit Commission which said that government support programmes are seen 'as a patchwork quilt of complexity and idiosyncrasy' (Audit Commission, 1989).

Aims of the Book

The book has two broad, interrelated aims. The first aim is to take stock of current British urban policy. By bringing together contributions from a range of experts on urban policy we hope to provide a reasonably up-to-date, authoritative and dispassionate analysis of the current situation. Subsequent chapters examine many of the initiatives that have been pursued under the banner of urban policy, not just in England but also in Wales and Scotland as well as abroad. We make no claim that the coverage is comprehensive. Inevitably there are omissions but most of the main dimensions of urban policy are addressed and where we have been unable to cover specific topics there are signposts into the available literature.

The second aim is to examine the current state of the art of urban policy evaluation. If urban policy is important, if it is contested and if it is rapidly changing where does this leave evaluation research? Critics argue that, despite the importance of urban policy, researchers whether in central or local government or in the academic community or elsewhere have not been up to the policy evaluation job. Certainly, as the book demonstrates in some detail, the challenges facing those concerned with urban policy evaluation are significant. But this is not to argue that it is an impossible task. Important research has been and is being done and it is clear that research efforts can provide helpful insights to those concerned with urban policy at all levels. However, there are important conceptual, methodological and political problems which have rarely been addressed in the literature and the book is an attempt to fill that gap.

In the remainder of this opening chapter we set the scene for subsequent chapters by offering a series of reflections on the current state of urban policy and the nature of urban policy evaluation in the 1990s. Naturally we hope that the arguments will be of interest to all those concerned with urban policy but it is also possible that the discussion of evaluation will be of interest to researchers working in other policy arenas.

The Origins of Urban Policy

Overviews of the evolution of UK urban policy are provided elsewhere and there is no need to repeat the full details here (Stewart, 1987; Lawless, 1988; Parkinson, 1989; Atkinson and Moon, 1994). It is, however, very useful to have an historical perspective in mind for, in several ways, current debates on urban policy rehearse arguments that have been debated before. It is just possible that by examining past experience we can identify lessons for future policy – or, perhaps more likely, uncover issues that have been shelved because they were felt to be inconvenient.

Whilst the origins of what might loosely be described as modern urban policy can be found in ideas germinating in the Home Office during the mid-1960s, the trigger for the introduction of the Urban Programme was Enoch Powell's infamous 'rivers of blood' speech delivered in April 1968. This heightened racial tensions particularly where black and Asian immigrants had settled in large numbers, mainly in the inner cities. The Urban Programme, announced by Harold Wilson, then Labour Prime Minister, in a major speech on immigration and race relations in May 1968, can be viewed as, at least in part, a measure designed to ameliorate the social conditions that might give rise to urban unrest. The aim of Urban Aid (as it was then called) was to help local authorities, by 75 per cent grant support of certain approved projects, to attack 'needle points' of deprivation.

The 1960s saw an upsurge in public concern about poverty and deprivation and campaigning groups, like the Child Poverty Action Group, put strong pressure on the Labour Government to pursue bolder approaches to a range of social policy issues, particularly housing. The contribution of community based campaigns and voluntary sector initiatives is often understated in explanations of the evolution of public policy. For example, the significant role of Shelter, the national campaign for the homeless, in shaping the early development of UK urban policy tends to be overlooked. The Shelter Neighbourhood Action Project (SNAP) involved setting up a small team to work for three years with very few resources in one of the most deprived areas of Merseyside. The idea was for the team to encounter urban deprivation on the ground, provide direct help to local people, demonstrate improvements and promote more effective policies for the future. The team's report was an imaginative analysis of the problems of Granby and Liverpool which contained bold proposals for a completely new kind of urban programme (Shelter Neighbourhood Action Project, 1972). This, one of the first serious pieces of urban policy evaluation, has stood the test of time very well.

The Conservative Party won the 1970 general election and Peter Walker, the new Secretary of State for the Environment, took a close interest in SNAP. He visited Granby in November 1970 and, when the project was completed in 1972, he recruited Des McConaghy, the SNAP director, to the Department of the Environment as a policy adviser. Here, then, we find the origins of a crucial shift in government thinking. The urban policy agenda was broadened partly because of the shift in leadership away from the Home Office to the Department of the Environment but, more significantly, because of the growing recognition of the interconnectedness of urban problems. At a seminar, held in September 1994 to celebrate the first twenty-one years of the School for Advanced Urban Studies at the University of Bristol, (the now Lord) Walker restated his belief in the importance of developing a 'total approach' to urban problems – 'the nearer you can get to a total approach the better'.

The 1970s witnessed a series of urban initiatives, some led by the Home Office (for example, the Community Development Projects and the Comprehensive Community Programmes), some forged by the Department of the Environment (notably the Inner Area Studies). In essence these various studies argued that the main cause of the 'urban crisis' was economic restructuring. *Inter alia*, the studies suggested that economic change had led to declining employment opportunities and a reduction in individual and community wealth in inner city areas. Peter Shore, then Secretary of State at the Department of the Environment, used these studies and other research evidence to underpin the launch of the Labour Government's inner city policy in 1977 (Her Majesty's Government, 1977).

In the early period of the inner city partnerships, created by the Inner Urban Areas Act, 1978, the pressure on local teams to formulate finite programmes, with their detailed lists of projects to fit fixed budgets and time-scales, distracted attention away from the development of more fundamental analytical work, policy evaluation and policy change (Hambleton, 1981). In short, there was a shift away from the idea of policy review and redirection towards an emphasis on the preparation and execution of inner area programmes. It is possible to argue that UK urban policy has never recovered from this shift. Urban initiatives, projects and programmes have proliferated in the years since 1979 but, as discussed below, there remains a continuing vacuum over the role and function of an urban policy.

The Conservative Approach to Urban Policy

After a period of uncertainty following the election of the Thatcher Government in May 1979 ministers announced their recommitment to inner cities 'policy' in September. Stewart (1987) notes that two strands of action emerged in the ensuing years. On the one hand, continuity and consolidation occurred – continuation of the partnership and programme arrangements, although with some 'streamlining' and with the addition of more local authorities to the list of programme authorities. A second, more important strand of activity, however, was the diversification of policy initiatives. Announcement has followed announcement with the result that, as mentioned earlier, urban 'policy' has become increasingly fragmented. Initiatives of the 1980s included Urban Development Corporations, Enterprise Zones, Task Force (Merseyside), the Financial Institutions Group, Urban Development Grant, City Action Teams,

Task Forces (Employment), Urban Regeneration Grant, National Garden Festivals, Business in the Community, Inner City Enterprise, Task Force (Confederation of British Industry), City Technology Colleges, Simplified Planning Zones, Safer Cities Schemes, City Grant, British Urban Development, Housing Action Trusts, Estate Action, the Valleys Initiative in South Wales, Training and Enterprise Councils, and more.

As well as being critical of this proliferation of initiatives the Audit Commission (1989) argued that the resources channelled by government towards inner city problems are 'relatively modest in scale'. Many in local government go further and argue that urban initiative allocations have been dwarfed by reductions in central government revenue support grant to urban local authorities.

A distinctive feature of the urban initiatives pursued by central government in the years since 1979 has been a desire to draw the private sector into the policy-making process. The 1988 policy statement, *Action for Cities*, was particularly pro the private sector – it suggested that the private sector should play the lead role in regenerating inner cities and stressed the theme 'helping businesses to succeed' (Her Majesty's Government, 1988).

Throughout the 1980s there was a continuing struggle between central and local government in relation to spending and policy priorities (Blunkett and Jackson, 1987; Lansley *et al.*, 1989). This conflict has led central government to bypass local authorities in developing many of its urban initiatives. The urban development corporations provide the most striking example of this strategy. The evidence shows that there has been a *massive* switch of government resources from the urban programme to the unelected development corporations in recent years (Colenutt and Tansley, 1990).

In November 1990 Margaret Thatcher was ousted as Prime Minister. On taking over the leadership John Major brought Michael Heseltine back into the government and, during his second spell as Secretary of State for the Environment, he appeared to press for a more conciliatory approach towards local government. Thus, Heseltine's announcement of the City Challenge initiative, in May 1991, suggested that Whitehall's anti-local government attitude had begun to change.

Certainly there was a strengthening of the local government role in that the statement invited fifteen local authorities to draw up programmes of action 'to tackle their key neighbourhoods'. Having said that the Secretary of State made it clear that he expected them 'to attract private finance and involve the private sector thoroughly in managing the programme' (Department of the Environment, 1991a). In the event eleven authorities were successful in this first round and each will receive £37.5 million over a five-year period. The City Challenge competition was rolled forward for a second year and, in July 1992, Michael Howard, the new Secretary of State for the Environment, announced a further twenty winning councils so that there are now thirty-one City Challenge authorities. On the one hand City Challenge has been welcomed by many of those involved in urban regeneration. Local authorities have been given a strategic role. The requirements relating to a comprehensive approach and community involvement are seen as steps forward. On the downside the competition for funds has created losers as well as winners and mainstream funding programmes remain heavily constrained (de Groot, 1992).

In July 1992 Michael Howard also unveiled plans for an Urban Regeneration Agency (URA) which, while it will focus on the 150,000 acres of vacant and derelict land in English cities, has the authority to work anywhere. The Agency, which was renamed English Partnerships in 1993, is a scaled down version of an earlier proposal, associated with Michael Heseltine, which envisaged a full-blown English Development Agency along the lines of the existing Welsh Development Agency. The impact of the English Partnerships' efforts are heavily dependent on levering private sector investment. The ability of the private sector to recover from the recent recession will, therefore, be a key factor in determining whether this initiative makes much impact.

A few months after the announcement of plans for the URA the Government made, in late 1992, three decisions which betrayed a serious lack of commitment to urban policy (Hambleton, 1993). First, it declared that expenditure on the Urban Programme was to be slashed. Second, the Government made it clear that there would be no third round of City Challenge in 1993. The third setback for urban policy, and this was one that received comparatively little publicity, was the Home Office announcement of cuts in grants under Section 11 of the Local Government Act 1966, a key source of funding to help ethnic minorities.

In response to criticisms that urban policy had all but collapsed the Government engaged in something of a rethink. There were three significant elements in the new package of measures announced by John Gummer, Secretary of State for the Environment, in November 1993: a Single Regeneration Budget (SRB), Integrated Regional Offices (IROs) and City Pride. The City Pride initiative is aimed at coalition building. In essence agencies in three major cities – London, Birmingham and Manchester – have been invited to put forward a vision for their area. This involves the various partners (the local authority, local business, the Training and Enterprise Council, the health authority, the police, educational institutions and other agencies) preparing a city prospectus covering the next ten years.

The SRB, which came into effect in April 1994, pulls together the resources from twenty different programmes. In round figures the budget is worth £1.4 billion in 1994/95. Because the SRB brings together a considerable number of programmes it can be claimed that the Government is responding to the criticism, mentioned earlier, that its programmes have become a patchwork quilt of complexity. By bringing different budgets together the SRB has the potential to develop a more coherent approach. It is also possible that central/local government relations will improve as the SRB provides a basis for dialogue and negotiation between local authorities and their regional offices. A major criticism of the SRB is that it masks a reduction of almost £300 million in the funds available – from around £1.6 billion for SRB equivalent programmes in 1992/93 to around £1.3 billion in 1996/97. A further concern relates to the fact that the original idea of each region preparing Annual Regeneration Statements was dropped in early 1994 following discussion in the Ministerial Committee for Regeneration. These statements would have set out the priorities for regeneration and economic development – critics argue that ditching the statements has created a strategic vacuum which makes the bidding process something of a hit and miss affair.

The IROs, set up in April 1994, bring together the existing regional offices of four departments: the Departments of Transport, Trade and Industry,

Employment and the Environment. They are intended to provide a single point of contact for local authorities, businesses and communities. A key task of the IROs is to administer the SRB although they also remain responsible for the other Departmental programmes operated by regional offices. The introduction of IROs represents a positive step towards better inter-departmental co-ordination within the Whitehall machine. Under the leadership of the new Senior Regional Directors the regional offices are attempting to provide a more responsive approach to the needs of different localities. It is too early to evaluate the changes but they do appear to have considerable promise. A major obstacle, however, is the long-standing functional division of power in Whitehall – this will not be transformed without considerable conflict. It is also the case that certain key departments – notably Education and the Home Office – are left outside the new arrangements. In this context it is possible that, despite the significant political differences, the English regions could learn something from the highly integrated approaches to regional management developed in the Welsh Office and the Scottish Office over many years.

It will be particularly interesting to see what happens to central/local relations as a result of the introduction of the SRB and the IROs. On the one hand it can be suggested that the arrangements might lead to a 'new localism' with more decisions being devolved to the regional level, thus creating new opportunities for local influence. On the other hand it can also be argued that Ministers have now become too accustomed to imposing their own priorities on particular localities. On this analysis the regional budgets and regional directors can be viewed as an extension of Whitehall control into the regions. In any event it would seem that the changes could well lead to a significant realignment in central/local relations (Stewart, 1994a).

The Urban Policy Debate

This discussion of the twists and turns in urban policy in the period since 1968 illustrates well the three themes introduced at the beginning of the chapter. First, the importance of urban policy is confirmed – it is unlikely that urban policy will disappear off the public policy agenda. There have been moments when it looked as if urban policy might be jettisoned – for example, in the months after the General Election in 1979 and in the winter of 1992/93 when Michael Howard was Secretary of State for the Environment. On both occasions, however, the Government saw fit to continue with at least some form of urban policy. This contrasts with the USA where, in the Reagan years, national urban policy was effectively abandoned.

Second, the discussion has shown how, over the years, urban policy has been reframed. Following Solesbury (1993) we can note that policy development often occurs within a prevailing frame or perspective. More radical policy development will require frame shifts which involve the setting aside of an accepted perspective in favour of a new way of organising and making sense of reality. It is a useful simplification to argue that urban policy has shifted from managerialism to entrepreneurialism – from provision of services and facilities to a concern for local economic development and employment growth (Harvey, 1989). This shift follows trends that have been established for many years in the

USA. As a result it is plausible to suggest that we have been witnessing a kind of Americanisation of urban policy (Hambleton, 1990a). This is not to suggest, however, that this is an inexorable trend. On the contrary there is now a growing body of cross-national comparative research which suggests that the regimes of governance in different countries, regions and cities reflect different histories and cultures. A consequence is that the space for local political initiative will vary from place to place (Keating, 1991). Even in the centralised state that the UK has now become it remains the case that, within limits, local authorities can innovate 'against the grain' of centrally imposed policies (Burns, Hambleton and Hoggett, 1994).

It is also the case that the opposition parties are mapping out rather different approaches to urban policy from the one being pursued by the Conservative central government. The Liberal Democrats in a recent policy paper stress the rejuvenation of communities and the empowerment of local people to change their surroundings (Liberal Democrats, 1994). The document argues that the crucial element in urban regeneration is the construction of a political framework within which local communities can exercise real power. Not surprisingly there is a focus on recreating urban democracy alongside measures to rebuild urban economies, improve the environment and enhance local services.

In 1993 the Labour Party set up an enquiry into urban policy chaired by Keith Vaz MP, the Shadow Minister for Local Government and Urban Areas. The enquiry, known as City 2020, spent eighteen months gathering evidence from cities across the UK and abroad. The emerging policy is driven by a desire to halt and reverse the economic and social polarisation found in so many cities. The new strategy, which is based upon boosting and supporting local initiatives and civic leadership, has five key principles: (1) a strategic approach going beyond the Single Regeneration Budget, (2) a strengthening of local democratic control, including local control of many of the quangos, (3) a decentralisation of power within cities to the neighbourhood or community level, (4) partnerships bringing together diverse interests, and (5) a new emphasis on community and voluntary organisations. Both the opposition parties, therefore, envisage a much less centralist urban policy than the Conservatives and see a much stronger role for the elected local authorities.

If we now turn to the third theme we can note that there is substantial evidence to support the claim that Ministerial announcements about new urban initiatives have proliferated. It is almost as if every Minister involved in urban policy over the decades has felt it important to announce measure after measure. At one level we should expect each Secretary of State to want to make his or her own distinctive contribution. After all the career of a Minister can be advanced by taking steps which give the appearance of action and which attract media interest. At another level, however, it does appear that urban policy has been peculiarly prone to 'the announcement syndrome'. It may not be too far fetched to suggest that the language about policies, rather than the policies themselves, has become the focus of attention (Edelman, 1977). On this analysis urban policy provides 'symbolic reassurance' that something is being done and is merely another example of the political viability of unsuccessful policies.

This brief overview of the evolution of urban policy and the current urban policy debate draws attention to a number of the challenges which face those

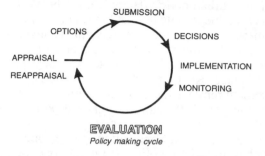

EVALUATION
Policy making cycle

Source: HM Treasury, 1988, p.19.

Figure 1.1   The rational policy-making cycle.

seeking to evaluate urban policy. There is widespread disagreement not only over urban policy objectives but also over the mechanisms for bringing about change. Moreover the objectives of urban policy are being constantly reshaped as different stakeholders gain advantage in the policy process. With some notable exceptions rigorous and systematic evaluation of urban policy has not been a dominant feature. At the same time the Single Regeneration Budget at £1.4 billion in 1994/95 represents a sizable government commitment. Inevitably questions have to be raised about how the effectiveness of urban policy is to be assessed and we now turn to this question.

Urban Policy Evaluation: the Current Debate

Either implicitly or explicitly, most discussion about how to undertake or use policy evaluation takes as its point of reference the role of evaluation in the rational policy-making cycle (see *Figure 1.1*). Without too much shoehorning of arguments into restricting categories, three kinds of reaction to this model appear to give rise to separate areas of debate, namely the *conceptual*, the *methodological* and the *political*. Of course, these are not watertight compartments, and the categorisations are drawn for ease of exposition.

Conceptual Issues

A fundamental conceptual difficulty with a positivist view of urban policy evaluation is the picture it presents of an integrated social reality, entirely external and independent of the observer/researcher about which ever-increasing amounts of knowledge can be accumulated. This view, by now discredited among social scientists, still has an extraordinarily powerful grip among agencies commissioning research for evaluation, who wish to ascertain the *facts*, as a basis for evaluation. Many commissioning agencies are unlikely to be impressed by researchers' concerns about the possibility of competing theories (often using incommensurable conceptual frameworks) of social and economic life. Nor are they likely to lose sleep over researchers' worries about the reflexivity of social research (and, indeed, social life) and the manner in which research is itself an intervention in social life which can influence perceptions, vocabularies and behaviour, thereby (wittingly or unwittingly) confirming (or invalidating) any crude 'natural science-style' predictions it might engage in.

In practice, many academic researchers accept the constraints imposed by the need to secure research funding. They work to a brief, and if this is narrowly defined, as it often is, they recognise that their research agenda is similarly restricted. Of course, it is open to researchers to use their contract research experience and findings to help generate alternative evaluation. Thus, Healey *et al.*'s (1988) appraisal of the British land use planning system – published as an independent evaluation – drew on a number of related research projects funded by a variety of sources, including central government departments. A practical difficulty with this approach is simply getting permission to re-use or re-work material collected as part of a research contract. Some commissioners of research try to retain control over research findings. Although researchers (and others) may object to this practice, as a matter of principle, one can understand the desire of agencies to control material which may have some kind of bearing on their future.

However, a more serious problem with this approach to urban policy evaluation is conceptual: the data collected as part of contract research may not be in a form which can be used in a straightforward way by an 'independent' evaluation. Data is identified and collected within a particular theoretical framework. A piece of contract research, though it may eschew overt theorising, will nevertheless be conducted within a particular theoretical approach. This will influence, among other things, the explanatory categories which are deemed to be useful, and the kinds of variables which will be observed. A different theoretical framework will generate different theoretical categories in its explanations, which will suggest different data gathering strategies for fieldwork. For example, Marxist and Weberian conceptions of class are, famously, different, and neither may make comfortable use of the Registrar-General's socio-economic categorisations based on occupation, which are used in so many pieces of contract research. We are not suggesting that it is futile to attempt to generate alternative analyses and evaluations of urban policy on the basis of what may often be limited contract research. But the difficulties need to be acknowledged, not least because they emphasise the intellectual importance of retaining some sources of funding for evaluation which are independent of agencies responsible for the implementation of urban policy. In the mid-1990s it is difficult to share, without qualification, Oatley's (1989) characterisation of universities as institutions which are relatively free from the constraints and pressures of central government sponsored research programmes. But he is wholly right in the importance he attaches to retaining a measure of financial independence, for it allows resistance to pressures to focus evaluation onto the immediate concerns of government agencies, and an exploration of the broader types of evaluation identified (for example) by Turok

(1991), such as those which question the very objectives of particular urban policy initiatives. Given the changes in the funding context the Economic and Social Research Council and the various research foundations are becoming increasingly important in fostering independent research on urban policy.

A further issue is the way the policy and implementation processes are theorised in the rational policy-making and evaluation model. It is by now widely recognised that public policies may have a variety of goals and objectives and that not all of these may be explicitly stated or even shared by those engaged in policy formulation or implementation (see, for example, Ham and Hill, 1993). The objectives of a programme may be contested, even as it is being implemented (Barrett and Fudge, 1981). Certain kinds of evaluation methodologies – notably action research – can be sensitive to, and indeed participate in, this process. But post hoc evaluations can be tempted to construct a misleadingly mechanistic and unidirectional model of policy objectives and implementation. Again, in some circumstances there may be implicit or explicit pressure on researchers to do precisely that, to 'sanitise' a messy process, and in so doing to exclude deviant perceptions and interpretations of its purpose and significance.

Methodological issues

The rational policy-making model rests on an implicit assumption that doing evaluation is unproblematic: the evaluation provides data which can straightforwardly inform future rounds of policy-making. The reality of evaluation is that the methodology has the potential to be so problematic as to even raise serious questions (on occasion) about whether it should feed into policy review and fine tuning at all. Consider, first, a case where conceptual issues shade into methodological ones, namely cost benefit analyses, or other evaluation methodologies which attempt to create a single scale against which to measure a variety of kinds of outcomes. The distortions introduced by attempting to convert all kinds of costs and benefits, including ones based on personal beliefs, values and upbringing, into a common currency, which will then be aggregated, have been well discussed (e.g. Sen, 1982; Williams, 1973, s. 7). Current best practice would advocate distinguishing between outcomes which are defined (and measured) along incommensurable scales (see, e.g., Friend and Hickling, 1977; Meyer, this volume). But, for outcomes which *are* measured along the same scale, there may be reasons for creating a single, aggregated, measure – for example, an index of some kind. The attractiveness of this approach to policy-makers is the apparent simplicity of dealing with a single measure or figure. But the simplicity conceals a complex of methodological issues and decisions, as Wong's contribution to this volume makes clear.

Approaches to evaluation of the kind favoured by the Audit Commission attempt to reduce performance measurement to a simple set of indicators – related, for example, to three Es (of economy, efficiency and effectiveness). Pluralistic evaluation, an approach which is being developed by policy researchers in various fields, is rather different as it attempts to take account of the *different* perceptions of success different interest groups will have. Such an approach is based on the understanding that there is a plurality of interests *within* the government machine as well as the range of sectional interests outside government. An adequate methodology needs to recognise that different interest groups will have different criteria of success and will attach different interpretations to the *same* outcome. An interesting example of this approach is provided by Smith and Cantley (1985) in their study of health care. Their book focuses on different 'meanings of success' in evaluating a day hospital – for example, free patient flow, clinical cure for patients, provision of an integrated service, beneficial impact on related services, support for relatives and service of high quality. Examples of pluralistic approaches to evaluation are found in Part Three of this book.

Three issues which appear consistently in reviews of evaluation methodologies (for example, Coulson, 1988, 1990; Gregory and Martin, 1988; Oatley, 1989; Turok, 1991) are: (1) assessment of additionality and indirect impacts, (2) the availability and quality of data, and (3) access to data. We consider each of these issues in turn.

The assessment of additionality or 'deadweight' is a key problem. In essence it revolves around the counterfactual question of what would have happened had a particular public policy (a project, a grant or whatever) not been implemented – although within this broadly defined issue there are a number of interrelated components which need to be identified and addressed (Pearce and Martin, 1994). To what extent has the policy contributed to particular outcomes (i.e. had an *additional* effect) and to what extent has it simply reinforced what would have happened anyway (i.e. been 'deadweight')? Such questions loom particularly large in evaluations of grant-giving regimes. They are fraught with the methodological problems which beset any attempt to isolate the effects of a single factor in multi-causal social relationships (see, for example, Sayer, 1992). But these are compounded by the desire on the part of clients of evaluative research of this kind not only to isolate a causal relationship but to quantify it. A linked problem relates to the assessment of the indirect impacts of a policy, as well as direct ones. Again, a key problem is the sheer complexity of causal linkages involved in socio-economic phenomena, and the difficulties of tracing the precise weight to be attributed to a particular intervention in a network of causation. As Gregory and Martin (1988) point out, this is no reason for ignoring indirect impacts, but particular caution needs to be used in both assessing and interpreting evaluations of them.

A second set of problems relates to the availability and quality of data. All too often the data best suited to addressing the research questions in a study may simply not exist. Standard publicly available data series – for example, unemployment statistics – may be useful, but generally will not be focused enough to provide the kind of data needed for confident evaluations of particular initiatives. It is better, therefore, if the data requirements of evaluation can be anticipated before studies are undertaken. For a number of reasons, this rarely happens. For example, it appears to be the case that most urban policy initiatives are not conceived and implemented with evaluation in mind. As a result little or no provision is made for considering data gathering/retention, and appropriate records are not kept, or are destroyed or mislaid. In addition the objectives of a policy may be so vague or multifarious as to defy comprehensive anticipation of how they might be evaluated. And, by definition, evaluations which seek to question the very rationale of an urban policy are unlikely to be ones which policy-makers and implementers are going to cater for. In these circumstances there is a danger that evaluation will be 'data-driven'. In other words what is evaluated and the nature of the questions asked will be unduly influenced by an awareness of the data which is available – evaluation may end up concentrating on what can be measured rather than what matters. An alternative approach would address such data limitations at the outset with a set of research questions being framed on the basis of policy imperatives and informed by an explicit conceptual framework (see Wong, chapter 8).

A further problem relates to the *quality* of data. In particular, to the extent that qualitative data (for example, attitudes, opinions, estimates) is collected from *stakeholders* (grant recipients, officials involved in implementation, policy advisers, etc.) then issues of *veracity* must arise (see Coulson, 1988). A recent evaluation of a number of urban policy grant regimes (Price Waterhouse, 1993) is an example of both the use of such sources, and how their limitations are overcome. The study was not unusual in its use of interviews with grant recipients in order to help calculate the likelihood of 'deadweight' or 'additionality'. But triangulation methods were employed – comparing data gathered by one method with that gathered by another – to restrict the likelihood of the study's being misled by reliance on any one data source.

The third set of problems concern access to data. Researchers engaged on studies which are independent of agencies implementing urban policies may have especially little leverage when it comes to gaining access to relevant data, much of which may not be in the public domain. For example, information relating to grants to individual firms is usually not available to the public and is exempted from public disclosure in the relevant legislation and government guidelines relating to open government (Cabinet Office, 1994 a, b). Consequently, evaluation research involving grants may need to rely on the co-operation of officers of the relevant agencies, and recipients of grants, for access to data, for interviews, and so on. Such co-operation is more likely if the researchers are deemed to share the values of the agencies they are evaluating (see Hambleton, 1994c) than if they are not (see Imrie and Thomas, this volume). Moreover, in an era where 'can do' and 'enterprise' have been valued politically above critical analysis and reflection it may be that many implementing agencies will view evaluation as an irrelevance or a diversion of energies from more important tasks.

Even researchers undertaking 'official' evaluation may not have untroubled access to relevant data. For example, as Coulson (1988) has remarked, the evaluator may be perceived as 'comrade, inquisitor or spy' (both within and outside the implementing agency), and such perceptions will influence reactions to her or him.

The politics of evaluation

Running through many of the debates and discussions labelled 'conceptual' or 'methodological' is a concern for the embeddedness of policy evaluation in systems of power relations operating at a variety of levels – societal (for example, class/gender based conflicts); institutional (for example, a central government department's relationship to a university); and individ-ual (for example, the pressure on individual researchers not to 'rock the boat' on a particular project even if s/he sees scholarly integrity being compromised by a desire or need to undertake the evaluation on the sponsor's terms).

Policy evaluation does not take place in a social vacuum: it is a social practice, or activity, and can bring gains (or losses) to individuals, groups and institutions. Imrie and Thomas, in chapter 9, speculate about whether the socio-political context of evaluation may not be beginning to impinge on researchers in increasingly direct ways. Crudely, the co-operation in research of agencies responsible for urban policy may come to depend upon what they see as 'in it for them', with the likelihood of criticism resulting in a refusal to assist. This scenario underlines the significance in urban policy evaluation of the co-operation of agencies who are themselves under the microscope – they are often repositories of useful data, as well as having a valid 'story' to tell from their own perspectives. In a discussion elsewhere of a particular research project, one of us has suggested that there might be particular sets of circumstances in which a co-operation rather than over-sensitivity between researchers and organisations is likely (Hambleton, 1994, p. 9):

> A local authority can't stop researchers from attending at least some of its meetings. This contrasts with decision-making in quangos which can be extremely secretive. The culture of the institutions being studied is almost certainly a key factor. In our research on local government decentralisation both the authorities were strongly committed to public debate and open government. As relatively healthy democratic institutions they had nothing to fear from scrutiny by researchers.

By contrast an institution which is relatively unused to being held to account in an open forum may find perfectly straightforward research activities rather threatening.

But the outcomes or conclusions of urban policy evaluation, and whether they constitute a clean bill of health for particular organisations, is only one focus for struggle and debate which is *political*, in the broadest sense. The issues we have raised earlier in this section also have such a significance. For example, the desire to safeguard capacities to conduct policy evaluation which questions the very assumptions on which the policy is founded, and perhaps does so within a theoretical framework not of the agency's (or government's) choosing, is politically loaded, in the sense that it is based upon a commitment to encouraging critical evaluation of state power. Again, arguments over the applicability or otherwise of methods such as cost benefit analysis are not simply 'political' in that their resolution can affect the nature of technical recommendations on policy, they can also reflect radically different views of what value is and of how it can be measured and judged. The epistemology of morals may not lead directly to particular party political positions, but it certainly involves different views of humanity's place in the universe, and these *are* significant in a broadly political sense. It makes a difference to those who are struggling for change, for example, whether environmental policy is based on utilitarian calculations or some kind of 'deep ecology' perspective, even if temporary coalitions between adherents of the two views are desirable (Beatley, 1994).

Returning to urban policy, Oatley (1989) and Turok (1991) have pointed out that the choice of *level* at which to undertake policy evaluation has political implications. What is allowed to be questioned by evaluation-related research can have important consequences for social relations. An extreme position, but one which has had its adherents in government from time to time, is to oppose *all* evaluation – and, hence, even the narrowest range of questioning. While this position may be justified by calling evaluation a diversion of energy and resources, as mentioned earlier, it also leaves the existing configuration of institutions and power relations unexamined and unchallenged.

It should be clear that far from being the routine, technical task envisaged by the rational policy-making model, policy evaluation raises a raft of issues of varying kinds. We would not suggest that this book (or, probably, any other single volume) can address all of them. However, we believe that the chapters which follow tackle some of the more important questions, and fulfil the aims of the book – to take stock of the current state of play in British urban policy, and to examine the state of the art in urban policy evaluation. The former task is largely achieved in Part One (which provides an overview), Part Three (which provides a selection of case studies) and Part Four (which allows comparisons between urban policy under different political orders). The latter task is addressed explicitly in Part Two, but contributors to other sections also consider methodological issues related to their work.

Outline of the Book[1]

Part One of the book sets British urban policy within an historical and institutional context, as well as providing evaluations of the overall effects of urban policy. The conclusion reached by Burton and Boddy, in chapter 2, is that urban policy has had little impact on the forces which have created an increasingly divided society. They base their view on a review of three aspects of the context of urban policy since the late 1970s. The first is the main social, demographic and economic changes that have taken place including long-term urban population loss, growing concentrations of poverty and rapid and extensive economic restructuring. The second aspect is political imperatives which have translated into policy, including the balance between growth and redistribution, the rise of property-led regeneration and the long-term reliance on area-based approaches. The final section of the chapter considers other aspects of policy context such as the changing role of local government and the emergence of new institutions, problems in the measurement of policy impact and the increasing significance of competition in the allocation of scarce urban policy resources.

In chapter 3, Bradford and Robson report some of the findings from the most comprehensive evaluation of urban policy in recent years. The aim of the research was to evaluate the overall impact of inner-city policies, not just the impact of individual policies. The authors make it plain that there were numerous methodological problems to confront, and it is interesting to note the constraints which the desired timetable of the research's sponsors imposed on how methodology evolved. Work was carried out on two major scales: the local authority level for 123 districts, and ward level for three northern conurbations. Work on the first was solely quantitative, while the second combined both extensive and intensive methods. The set of objectives of policies (explicit or implicit) was collected from government documents and encapsulated under ten major headings. Two over-arching aims summarised these ten: to increase employment opportunities and to improve the area's attraction as a place to live. This process informed the choice of outcome indicators used to examine the impact of government inputs in the quantitative analysis. The results of these quantitative analyses are reviewed alongside the major findings of the qualitative work, to present an overall view of successes and failures of urban policy in the 1980s.

In chapter 4, Stewart puts the spotlight on the management of urban policy expenditure evaluated against the substantive objectives to which the instruments of policy are addressed. Financial resources are usually seen as an input to, rather than an output from, policy intervention. However, the containment of public expenditure has, itself, been a key public policy objective in recent years. His analysis maps the shifts in actual and planned urban policy public expenditure over the period 1988/89 to 1996/97 and examines what is happening to urban expenditure as a result of the introduction of the Single Regeneration Budget in 1994. He shows how urban expenditure has been vulnerable to the impact of cyclical movements in the development market, that expenditure has been volatile and unpredictable and that the Single Regeneration Budget involves a cut of almost £300 million in the funds available between 1992/93 and 1996/97.

It is increasingly unrealistic to consider British urban policy without reference to the European Union. Chapter 5 presents a complex and dynamic picture: the formal encroachment of 'Europe' on urban policy-making remains limited, but in an era of urban entrepreneurialism any organisation which has funds is the target of (and influential among) urban policy-makers. The ubiquity of signs noting the contribution of the European Regional Development Fund to infrastructure projects in Britain's 'peripheral' regions makes clear the potential leverage which the European Union has in steering national urban policy in particular directions. Chapman argues that there is evidence that there is a nascent programme of urban policy in the pursuit of which this leverage could be exerted.

Part Two of the book considers a series of methodological issues, chosen to illustrate the wide range of issues which confront urban policy evaluation. In chapter 6 Meyer explores the problems, for evaluation, of differences in perceptions of various constituencies with an interest, or stake, in urban policy. Beginning with the evaluation problem and a general overview of requirements for *consistency* and the problems they generate in policy comparisons, it turns to evaluation objectives and conditions for *accuracy*. Some case studies drawn from the UK and the USA are considered, before it examines the problem of responding coherently to the impossibility of objectivity in the context of a high demand for some form of quantitative evaluation and assessment of 'value for money'.

If Meyer's chapter considers methodological problems at a level of some generality, in chapter 7 Martin and Pearce focus on a very specific, but important, example of urban policy evaluation practice – the appraisal of proposals for funding. Their chapter investigates the appraisal systems and practices applied in four of the UK government's key urban policy initiatives which were operating in the late 1980s: the Urban Development Corporations, Derelict Land Grant, City Grant and the Urban Programme. It highlights, and seeks to explain, the variability and shortcomings of appraisal practice and assesses the ways in which procedures might be improved. Two major problems are highlighted. Firstly, systems failed to address adequately a range of key, technical issues. Secondly, they largely lacked mechanisms for assessing the wider, strategic dimension in which individual projects were implemented. The chapter concludes that, despite recent action by central government designed to remedy these faults, there is a continuing need for the enhancement of the technical aspects of appraisal and, more fundamentally, for the adoption of a strategic approach. Failure to address these issues threatens to undermine the effectiveness of inner city initiatives.

Urban and regional policies are increasingly applied on an area targeting basis, thus increasing the need for improved information on the candidate areas and better methodologies to aid the prioritisation process. The pressure for more and better information is also increased by recent academic and policy debates on the importance of identifying the distinctiveness of individual regions and localities in terms of their strengths and weaknesses. In chapter 8, Wong identifies four basic steps for a well-founded targeting analysis: clarifying the concept measured; specifying the key issues by which the concept is to be represented; identifying adequate statistical indicators covering those issues; and creating an overall index to summarise these indicators. The latter part of the paper emphasises that there are numerous ways to produce a multi-variate index; however, the choice between them is not simple and will greatly affect the results obtained. Although the chapter stresses that different options will be more appropriate for different purposes, some 'best practice' points are identified.

In chapter 9, Imrie and Thomas consider the implications for urban policy evaluation of trends in urban governance. They use a case study to illustrate the argument that there are trends within urban governance which are fostering values antithetical to external scrutiny, and creating conditions where independent evaluation may be difficult.

Part Three provides further examples of urban policy evaluation in practice, but in contrast to the broad canvas of Part One, these chapters focus on qualitative analyses of individual case studies. All three chapters consider the highly topical question of community governance, and its implications for urban policy formulation and delivery, and all three take seriously the methodological and conceptual implications of pluralistic evaluation. In chapters 10 and 11 the City Challenge initiative is analysed. After the top-down, and often insensitive, approach to urban policy of the urban development corporations, City Challenge was touted as offering a real partnership for local communities and local authorities. However, not all those who bid for City Challenge funding from central government were successful, and Oatley and Lambert are surely right to argue that an evaluation of the initiative as a whole should take into account the perspectives and experiences of unsuccessful bidders.

Their chapter uses results from a questionnaire survey of all authorities that bid unsuccessfully in City Challenge and an in-depth case study of one unsuccessful authority (Bristol) to analyse the impact of City Challenge on those authorities that did not secure funding and to assess the implications of the widespread application of competitive bidding to the allocation of urban funding. They conclude that the bidding process itself wrought some changes in institutional relationships, but that failure to secure monies has serious repercussions for areas which are often demoralised by social, economic and political marginalisation.

Davoudi and Healey's chapter (11) explores the processes of governance within an area which was successful under City Challenge. The chapter sets the objective of the initiative to involve residents of areas of disadvantage in the context of contemporary conditions in such neighbourhoods. It then assesses the experience of the early stages of two City Challenge programmes in order to assess the character of the 'partnerships' which are evolving. The two examples vary significantly in their style and in the strategy of the local authority, as well as in the characteristics of the areas. It is argued that while central government's influence over the style of the programmes remains pervasive, the strategy of the local authority and the struggles by the participants in the partnerships have a significant influence on the content and style of the programmes. Further, the style of the programmes affects the balance of power within the programme and the terms of 'incorporation' of the various partnerships in governance processes.

Variation is also a theme running through Hastings and McArthur's evaluation, in chapter 12, of community involvement in regeneration strategies in urban Scotland. Their account is valuable because it analyses initiatives which have been underway for some years: the dynamic of 'community partnership', and its changing meanings over time, emerge quite clearly, as do different interpretations of objectives and 'reality'. Though their chapter ends with a series of new questions, it provides a sensitive analysis of activity to date which will be useful to those involved in newer initiatives such as City Challenge.

In chapter 13 Morgan examines the Valleys Programme and governance structures in Wales. The chapter introduces a useful comparative dimension to the book. South Wales never embraced the free-market credo of Thatcherism and Morgan argues that the Programme for the Valleys represents a form of partnership approach. He offers a critique of the programme which suggests that the resources allocated were inadequate and that the initiative suffered from significant design and delivery weaknesses. His examination of governance structures suggests that the new Integrated Regional Offices in the English regions would benefit from studying the strengths and weaknesses of the Welsh experience – not just the work of the Welsh Office, but also the urban joint-ventures developed by the Welsh Development Agency and the local authorities.

Green and Booth extend the comparative theme by offering an appraisal of French urban policy in chapter 14. The chapter discusses the Contrat de Ville at a conceptual level before presenting some initial observations about its implementation in a case study of the Communauté Urbaine de Lille. The chapter notes the essentially social nature of the Contrat or contract, the innovative nature of intercommunality and the emphasis on partnership. It highlights the complexity of the process as well as the concern for solidarity or pulling together. The authors argue that it is useful to study the French experience as it provokes fresh thinking about practice in the UK.

Anglo-American dialogue in the field of urban policy has intensified in recent years. The reinventing government debate, which has gathered pace on both sides of the Atlantic, suggests that reshaping the institutions of city government and city governance is likely to be critical to the success of future urban policy. In chapter 15 Hambleton argues that, whilst it is useful to compare and contrast the national urban policies emanating from Washington and Whitehall, it is even more rewarding to engage in transatlantic comparison of systems of city governance. The chapter provides an examination of US models – the mayor-council form and the council-manager form – and, through case studies of Baltimore, Maryland and Phoenix, Arizona, the chapter highlights the possibilities for further transatlantic policy transfer in the sphere of urban governance.

Note

1. In summarising chapter contents we have sometimes drawn on and used abstracts supplied by authors, for which we are grateful.

PART ONE

Urban Policy in Context

2

THE CHANGING CONTEXT FOR BRITISH URBAN POLICY

Paul Burton and Martin Boddy

Introduction

In 1977 the White Paper *Policy for the Inner Cities* (DoE, 1977) set out what had become the conventional wisdom on the nature of the inner city problem by focusing on the interlocking aspects of economic decline, physical decay and social disadvantage. Although subsequent Conservative governments have added to this diagnostic list anti-growth sentiments among local authorities, the growth of single person households and lack of entrepreneurial spirit, it still provides a useful benchmark against which to measure change since then.

Looking at the 1977 benchmark in some more detail we see that economic decline included a systematically higher rate of unemployment among inner city residents, a mismatch between skills and available jobs and a general lack of demand for labour in some major cities such as Glasgow and Liverpool. The failure of modern service sector growth to make up for the long-term decline in manufacturing employment was acknowledged, as well as the loss of jobs in the more traditional service industries of the docks and railways. The White Paper also recognised the persistence of a shabby physical environment which served to make many inner areas unattractive to residents and prospective investors. The existence of large areas of vacant land was also noted, along with the opportunity this presented to tackle many problems, including the need for public open space. In terms of social disadvantage it was argued that a collective form of deprivation affected all inner area residents even though individually they might have satisfactory homes and worthwhile jobs. This was in turn used to support a policy of 'discriminating in favour of the inner areas in the working out of public policies and programmes' although the possibility of a higher scale of welfare benefits for inner area residents was considered untenable.

How then have conditions changed since the White Paper was published, and what else can we say about the changing context for urban policy over the last three decades? This chapter tackles these questions in three parts: the first discusses some of the main social, economic and demographic changes that have taken place over the last few decades and looks at their impact on urban and inner city areas; the second part discusses some of the key issues facing policy-makers in the light of these findings; while the final part considers broader contextual issues in the development of urban policy both past and present.

Economic, Social and Demographic Context

The well-being of cities and the people who live in them is tied closely to the state of the national economy as a whole, and the recent recession, like those of previous decades, has had a profound effect on the opportunities available to people in cities. Although middle-class households in the affluent suburban south-east have suffered more under this recession than previously, people living in the large conurbations and older industrial areas have continued to bear the brunt of economic change and decline. While nationally there has been a degree of economic diversification and tertiary sector growth, the large cities have retained a disproportionate share of those industries most vulnerable to change and decline.

The future prospects for the economy as a whole are, of course, difficult if not impossible to predict or even to forecast. We can, however, identify a number of elements of economic change and restructuring which have been significant in recent years and look set to remain so in the medium term. Economic growth without an equivalent increase in employment growth will continue to be an important feature, and hence unemployment is likely to remain high. High levels of unemployment in the core urban areas are likely to persist given a number of factors: the level of competition for jobs (Begg, Moore and Rhodes, 1986); the changing nature of demand for labour; and employers' recruitment strategies and views on employability (Davies and Mason, 1986). Those living in inner city areas and old industrial cities will tend to benefit least and latest from any general increase in the demand for labour associated with a recovery.

In terms of industrial change inner city areas remain particularly vulnerable. The situation varies between places, but many core urban areas will continue to be especially prone to continued decline in employment over the next five to ten years because of their reliance on declining industrial sectors and older, less productive plant. Slower growth, restructuring and job loss in the service sector will also be an increasingly important factor. This is likely to impact disproportionately on inner city workers who tend to be concentrated in those segments of the service sector labour force most vulnerable to change.

Linked to this, forecast changes in occupational structure and in particular the declining demand for semi-skilled and unskilled labour will have adverse impacts on urban labour forces. This will be reinforced by technological change, more rapid turnover of skills and a general rise in skill requirements in both manufacturing and service sectors. These various changes will further disadvantage inner city residents who will increasingly be confined to less stable, low-wage segments of the job market with few opportunities to participate in good quality training schemes.

The so-called 'demographic dip', the drop in the numbers of young people coming on to the labour market, will not have much impact on inner city areas because of the existing over-representation of younger groups in these areas. Correspondingly, even if labour markets tighten up as a result of economic recovery the impact of demographic effects will be least pronounced precisely in high unemployment inner city areas.

In the face of these economic changes a number of important socio-demographic developments and trends are evident. These include the rise in smaller and single person households, the growth of the elderly population especially the very old, an increasingly difficult transition to adulthood for many young people and persistent racialised segregation.

There is likely to be a long-term and steady increase in the number of smaller (i.e. one and two person) households over the coming three decades simply because of changes in the age distribution of the population (Ermisch, 1990). In addition the effects of growing rates of cohabitation and the continued impact of divorce and separation will fuel the long-term growth of smaller households. The consequence is likely to be an increased demand for housing and in particular for smaller units and while elderly people are perhaps more likely to prefer settings outside the major cities, younger households may demand new housing within the conurbations.

In spite of this potential development there has been a long-term decline in the population of the inner areas. Begg, Moore and Rhodes (1986) chart this decline since the 1950s by looking at the population of working age in different types of area (see Table 2.1).

Table 2.1 Changes in the population of working age by type of area 1951–81

Type of area	1951	1961	1971	1981
Inner cities	100	93	77	65
Outer cities	100	92	87	87
Free standing cities	100	96	94	91
Towns and rural areas	100	106	114	117

Note: GB = 100 in 1951.
Source: Begg, Moore and Rhodes (1986).

It is important to sound a note of caution in making or utilising any long-term population projections, for external shocks can and do have significant impacts on the assumptions built into population models. Champion (1992) describes how the phenomenon of counter-urbanisation so apparent for the 1970s in Britain appears to have been superseded by a return to more traditional patterns of population distribution in the 1990s and concludes by saying that 'the rapid speed of recent demographic change, the wide diversity of its outcomes and the high degree of interrelationships with other aspects of urban and regional change all introduce huge uncertainties into the task of plotting out future trends' (p. 477).

Willmott's (1994) comparison of population change in 'deprived areas' between 1981 and 1991 reveals a similar pattern of loss in all inner areas except Tower Hamlets in London, where the high fertility rates among the relatively youthful Bangladeshi population and a degree of in-migration to new housing in the Isle of Dogs resulted in a 7.5 per cent population increase over the period.

This case illustrates another significant contextual aspect of social change in urban areas, the persistence of residential segregation along ethnic or 'racial' lines. Acknowledging but not entering into the debate on the construction, meaning and relationship between 'race' and ethnicity, we can say that by 1991 approximately 6 per cent of the total population of Britain belonged to a minority ethnic group. Given the fertility implications of this relatively young population as a whole, it is likely that the total minority population will increase to around 10 per cent of the national total over the coming decades (Ballard and Kalra, 1994).

Unlike the white majority population, minority ethnic groups are highly concentrated in terms of where they live. For example, while 23 per cent of the total 'white' population live in the large metropolitan areas of Britain, over 79 per cent of Black Caribbeans, 65 per cent of Indians and Pakistanis and 74 per cent of Bangladeshis do so. At the local authority scale the degree of concentration is even more pronounced as only 29 districts (6 per cent of the total in Britain) account for 55 per cent of the total minority population and 22 of these are London boroughs. Furthermore, within districts we know that people belonging to minority ethnic groups are concentrated into a narrow range of wards and enumeration districts (Burton and Stewart, 1995). Of course this might be of little significance if the pattern and degree of concentration was simply a reflection of the wishes of the people in question, but as Smith (1989) observes it is difficult to understand why the areas they have chosen to remain concentrated within should typically be those with poor quality housing and poor job opportunities. It is more likely that concentration and segregation along ethnic or 'racial' lines is a product of racism and discrimination in both housing and labour markets, whereby black people are trapped in areas of declining job prospects by institutionalised inequalities in housing markets. There are few signs that people from minority ethnic groups have become more dispersed in their patterns of residency and in so far as this has been an important but implicit element of urban policy since the 1960s, it is also a significant indication of urban policy failure.

Another area of policy failure would seem to lie in the increasingly difficult transition to adulthood faced by many young people. Although the proportion of young people entering higher education expanded significantly during the 1980s and early 1990s, public expenditure constraints are now resulting in a tailing off of this expansion. However, for the many young people for whom this is an implausible option, traditional routes into the labour market have been substantially reduced at the same time as a wide range of welfare benefits have been restricted in an attempt to encourage young people to accept wages and conditions more in line with their disadvantageous position. The transition to adulthood for this group has, therefore, been made more difficult and many young people now find themselves stranded without many of the material trappings of adulthood and with few prospects of independent living.

While the causal connections between economic disadvantage and crime remain subject to intense political debate, it is clear that young men have played a major part in the significant increase in recorded crime over the last decade, especially in the fields of car crime and burglary. It is also clear that the incidence of reported crime is distributed very unevenly over the country as a whole, with much higher rates in those areas targeted by urban policy namely the inner areas and peripheral estates. Table 2.2 illustrates this pattern using indices compiled from the last three British Crime Surveys (1984, 1988, 1992) applied to ACORN neighbourhood groups. These ACORN groupings represent a system for classifying households according to various demographic, employment and housing characteristics of their immediate neighbourhood and are increasingly used in both commercial market research and public policy development.

Table 2.2 Comparative crime rates by type of area 1984–92

ACORN area	Burglary	Autocrime	Robbery
Agricultural	20	20	50
Modern family housing	60	70	70
Older housing	70	100	60
Affluent suburban	70	70	70
Retirement areas	70	80	70
Older terraced	120	160	100
Better off council estates	90	110	120
Less well off estates	150	160	100
Poorest council estates	280	240	200
Mixed inner metropolitan	180	190	340
High status non-family	220	150	250
Indexed national average	100	100	100

Note : Draws on data from British Crime Surveys carried out in 1984, 1988 and 1992.
Source : Mayhew and Maung (1992).

Finally in this section we turn to more general points in the geography of disadvantage which provide further context for the analysis of urban policy.

There was something of a debate from the mid-1980s onwards as to whether the outer estates of most cities suffered worse levels of deprivation than their corresponding inner city areas, or more generally whether deprivation was greater on the outer estates than within inner city areas. While it is clearly important to know with as much precision as possible where the areas of greatest deprivation are located, the argument between two poorly defined types of area as to which is worst off has not been fruitful (Perry, 1991; CES, 1985).

As Thake and Staubach note, multiply deprived neighbourhoods can be found in inner city areas, at the edge of cities and in locations in between (1993, p. 20). Location in this typological sense is not a determinant of deprivation any more than tenure or ethnicity; however, it might help to explain the particular configuration of forces which creates a concentration of deprivation in any particular locality.

De-urbanisation and counter-urbanisation look set to continue, with population and employment decline in the major conurbations and larger urban areas combined with growth in smaller towns and more rural areas. Crucially, past experience suggests that the selective nature of this decline leaves behind it an increasing concentration of those disadvantaged in labour and housing markets within the core urban areas. New employment growth, meanwhile, is concentrated in fringe conurbation areas, in smaller provincial towns and in rural areas. Those better placed in the labour market increasingly commute back to the core urban areas and represent strong competition for inner city residents for city centre jobs.

On current trends, there are likely to be increasing numbers of households falling below the commonly used poverty thresholds and growing divergence of earnings, in particular for the bottom decile of the population. There is an increasing concentration in inner city areas of 'the new poor' – lone parents, the old and very old, the economically marginalised young, the long-term unemployed and the prematurely retired.

Willmott (1994) and Willmott and Hutchinson (1992) have compared living conditions and life opportunities in some of the most deprived parts of the country with a variety of other areas – other deprived areas, surrounding regions and more prosperous areas. They focus on local authority districts, primarily because of limitations on the availability of data at other spatial scales, and select the 36 most deprived districts from Great Britain using the Department of the Environment's index of urban deprivation for 1981 and index of local conditions for 1991.

The list of the most deprived districts in England varies to some extent between 1981 and 1991 as six districts moved on to and off the list. The majority of areas in this 'most deprived' group have, however, remained so over the decade of the 1980s while those that moved out of the category did not move a lot further down the ranking of all districts – the lowest ranking of these being Burnley which moved from 27 in 1981 to 56 in 1991. However, it must be remembered that the indices used for each point in time were constructed using different variables and different assumptions about the ways in which they combine.

Green (1994) has recently analysed the changing geography of poverty and wealth using data from 1981 and 1991 and identified a clear pattern of growing polarisation between the 'best' and 'worst' areas at a ward level. She too uses three alternative measures of distribution – the degree, the extent and the intensity of poverty and wealth within particular localities or areas – and identifies a greater concentration of poverty over the 1980s. Moreover there has been an increase in the number of areas exhibiting multiple aspects of poverty or 'concentrated poverty' in the large urban centres, particularly London, and the older industrial areas. London experienced a growth in areas of concentrated poverty and concentrated wealth and as a consequence its overall social structure became more polarised. It is also worth noting her finding that minority ethnic groups, especially those identified by the Census as being Bangladeshi, Pakistani or Black, are concentrated in areas of extreme deprivation and became more so over the decade.

Putting this all together, one can only conclude that there is little to suggest that inner city problems have diminished or are likely to. On the contrary, there is every indication that, based on current trends, the future for inner city areas in Britain is one of increasing spatial concentration of unemployment and disadvantage, increasing polarisation, and increasing breakdown of social and economic cohesion within inner city areas and between inner cities and other areas.

It is worth remembering that British cities do not exist in splendid or even ignominious isolation, but are part of a constellation of European cities affected by global forces of social and economic change. Cheshire's composite indicator of the long-term and structural problems of functional urban regions (FURs) within the member states of the EU gives a useful picture of the position of British cities within this constellation (Cheshire, 1990; Cheshire, Carbonaro and Hay, 1986). He draws a broad distinction between two main groups of FURs facing major difficulties – the growing cities of the south of Europe and those running in a band from north-east Italy, through eastern France, the Ruhr and southern Belgium through the north-west of Britain to Glasgow and Belfast. This latter group have tended to experience population decline and problems associated with their reliance on older declining industries. Moreover, there has been a growing polarisation between the best and the worst cities, the worst have tended to decline relative to those in the strongest position. It is also worth noting Cheshire's conclusion that urban policy can have a marginal influence on the performance of cities or FURs 'although most of urban performance seems to be determined by factors over which policy can have no influence' (p. 332).

It is also worth remembering that spatially concentrated deprivation in core urban areas reflects the combined effects of the labour market processes and poverty on the one hand and housing allocation processes, both market-based and bureaucratic on the other. By and large, housing allocation processes concentrate and trap those lacking financial and other resources in the poorest quality housing stock in the poorest quality urban environments. Many if not most of those who have the resources to leave, do so. Moreover, the continued vulnerability of such areas to economic restructuring and job loss, together with the effect of counter-urbanisation, makes the situation even worse.

We make this distinction because labour market disadvantage is clearly *one* key determinant of inner city disadvantage. The concentration of those less competitive in the labour market in areas of high unemployment and strong competition for jobs further increases that disadvantage. In other words it is *harder* for inner city residents to get jobs than the unemployed elsewhere. However, labour market disadvantage is not the only dimension to the problem. For other groups, the 'new poor' that fall below commonly used poverty thresholds – pensioners, lone parents, the sick and disabled – benefit levels and eligibility criteria on the one hand and direct state provision of services and facilities on the other are key determinants of their quality of life. Despite a steadily growing welfare budget, there has been an increase in the use of targeting and means testing in the application of specific welfare benefits. The recipients of many benefits have therefore found themselves confronting welfare regimes of increasing toughness. Linked with this development has been an ongoing debate of great policy significance concerning the existence and causes of an underclass. While most academics have been highly critical of the application of the notion in Britain (Morris, 1993), the spectre of wholesale welfare dependency in some neighbourhoods and its consequences for local social relations has been used to justify these more general moves to toughen welfare regimes (Gans, 1993).

It is worth rehearsing these definitional points about inner city areas and the causes of inner city problems because of the way they provide the context for the development of policy measures to tackle the problem. In short we have highlighted the importance of processes of change and the management of change – in labour markets, housing markets and welfare systems – rather than the characteristics of defined inner city areas. The construction of an inner city problematic in terms of the characteristics of its victims has, however, dominated the urban policy debate at various times over the last three decades and continues to hold out the prospect of quick-fix policy solutions to hard-pressed politicians. We move on in the next section to look at some of these policy issues in more detail.

Policy Issues

Looking at the combined effects of two decades of specific urban policy measures, as well as the impact of the main spending programmes of government and fluctuations in the economic fortunes of the country, we find it difficult to conclude that the problems faced by inner city areas have been dealt with successfully. The overall picture is one of very little change – certain parts of large towns and cities still experience high levels of unemployment, job opportunities are still extremely limited, housing conditions continue to be poor and public services offer little by way of compensation. Moreover, crime and fear of crime is growing, racial discrimination continues to find new expressions in many different institutional settings and local political and community organisation remains fragile and vulnerable. Chapter 3 (by Bradford and Robson) deals in more detail with measuring the impact of urban policy over the last two decades, but the overall conclusion remains one of limited impact on the quality of life and opportunities for inner city residents.

At one level this is very depressing although it also suggests that it would be foolish to expect policy miracles in the face of such deeply ingrained economic and social processes. However, it is clearly important to assess critically the policy alternatives. There are real choices to be made between policy alternatives for the future, and lessons to be learned from the past, and so we move on now to consider a number of issues facing policy-makers based on past experience and future trends.

Economic policy is not labour market policy

Not only is the very notion of distinct inner city economies questionable, but strengthening local economies by attracting investment and creating jobs in or adjacent to high unemployment areas may have little impact on those living in the inner city. The juxtaposition of wards with the highest unemployment rates close by the main concentration of retail and office-based employment in most of our larger urban areas demonstrates this. In short, economic policy is not labour market policy and economic growth is not enough. At one level it is even irrelevant – many jobs are already there. The great majority of vacancies or job opportunities are created by turnover in the labour market as people leave existing posts for new jobs, to retire or to take up full-time domestic labour. It is a question of who gets the vacancies thus created which is at issue.

The need for economic growth and job creation before inner city unemployment can be addressed is often an excuse used to justify and to maintain unequal access to jobs and the existing distribution of employment and labour market opportunities. Explicit mechanisms are needed to link those disadvantaged in the labour market to opportunities for employment and good quality training (Boddy, 1992). Employers' recruitment and selection strategies are often the main barrier, more so than any lack of skills, training or relevant experience on the part of the inner city workforce. Here, therefore, it is the policies of the Employment Service, the Department of Employment, Guidance Services and in particular Training and Enterprise Councils which have the key role to play – rather than the Department of the Environment which retains the key role in official *urban* policy.

Quality of life and the dominance of the economic

Strengthening local economies, job creation, training and investment are usually the dominant themes of urban policy. These are self-evidently important to those marginalised in the labour market, those seeking work and those dependent on them. But for many inner city residents including increasing numbers of the 'new poor' referred to earlier, the economy and the job market is largely irrelevant as they are no longer members of the labour force.

Changes in benefit levels, in eligibility criteria and in taxation can have much more profound effects than marginal shifts in urban policy. Benefits for single parents, retirement benefit, value added tax (VAT) on domestic fuel, housing benefit and so on are often more significant in affecting the quality of life of these inner city residents. These groups of people are disproportionately reliant as well on the public sector for the direct provision of many services and have suffered most from the cutbacks in public sector housing provision, local authority spending and welfare provision over recent years.

As both this and the previous point makes clear, the policy context set by other government departments, impacting on benefit levels, employment and training have potentially much greater implications for inner cities than many urban policy measures introduced by the Department of the Environment.

Development-based approaches and property-led regeneration

There is now ample evidence that physical development and urban renewal may improve the urban fabric and the physical environment with little if any benefit to existing inner city residents, the lowest income groups and most disadvantaged (Davidson, 1976). This is at the heart of the difference between conflicting definitions of urban problems and policy paradigms. Physical and environmental projects and development-based activities have a place but they need to be closely integrated with employment and training, with people-based solutions and with community involvement if they are to counter rather than increase marginalisation and exclusion.

Area-based initiatives

Area-based initiatives and spatial targeting have long been a central feature of urban policy (Burton and O'Toole, 1993). However, City Challenge and the Single Regeneration Budget (SRB) represent a point of departure, in that they combine spatial targeting with competitive selection, unlike previous approaches which usually allocated resources according to formulae designed to reflect local needs. Thus, the spatial concentration of deprivation has been the main way in which the targets of urban policy measures have been identified and concentration has been used both to select areas on the basis of the intensity of their problems and to maximise the potential impact of local interventions.

Although the locality studies carried out under the auspices of the Economic and Social Research Council (ESRC) pointed to the value of careful and detailed research for understanding the dynamics of local socio-economic change and development, we still do not fully understand the mechanisms by which elements of deprivation reinforce each other within particular areas (Stewart, 1988). Moreover, we are still faced with the paradox of targeting whereby 'those small residual pockets of deprivation so attractive to policy strategists turn out to be neither small nor residual' (Deakin and Edwards, 1993, p. 69) and the suspicion remains that spatial targeting is driven as much by the need for resource rationing as it is by any clear analysis of local circumstances.

Spatial targeting is not enough

While they are likely to remain important features of urban policy, area-based initiatives and spatial targeting are not enough. Mechanisms are needed to ensure that the benefits of policy impact on target groups within the labour force at large and on relevant households and individuals. Improvement of housing or the environment on an area basis will, for example, be counter-productive if this leads to displacement of the existing population or gentrification. We need to bear in mind though that to some policy-makers and analysts, displacement or the dispersal of 'problem households' represents a desirable rather than undesirable feature of urban policy (e.g. Fothergill, 1988; Hall, 1990). Moreover, the creation of new jobs or training places in inner city areas will, equally, fail to the extent that such jobs and training places are taken by those from outside the areas. Linkage or targeting at the level of individuals not simply areas must, therefore, become a key criterion against which to judge future policy initiatives. Figures claiming so many jobs created, so many housing units improved or so many training places funded are meaningless unless we know who benefited from them and what were the knock-on effects.

Market-based approaches and the enterprise solution

Finally, there has been increasing reliance in urban policy on market-based approaches – levering-in private sector resources, involving business in the development of local policy and introducing a more competitive edge to the distribution of resources. There is a contradiction here as government has sought to use market mechanisms to solve the problems created by market failure such that 'the contradictions of policy bear heavily on urban problems, exacerbating the normal problems of uneven development and lagging regions' (Moore, 1993, p. 234).

As the most recent period has demonstrated, however, market-based approaches and the enterprise solution are heavily dependent on the overall performance of the economy and are least able to address issues during recessionary periods when inner city areas are hardest hit. In addition attempts at exploiting the social responsibility of industry have singularly failed and reliance on the benevolence of developers offers only marginal benefits such as 'improved shopping facilities, at best; or a fistful of small grants from a community trust and some ornamental statuary to help locals feel better about being poor' (Deakin and Edwards, 1993, p. 255).

Policy Context

So far we have described some of the key social, demographic and economic trends affecting inner city areas and the particular policy issues facing government in its attempt to do something about the problems faced by people in inner city areas. This final section concludes by identifying the other main contextual factors affecting the development of urban or inner city policy.

The changing role of local government

From the time of the 1977 White Paper to the present day, the role of local government in the regeneration of urban areas has fluctuated markedly – at least in the eyes of central government. In 1977 it was to be one half of the new partnership between central and local government designed to tackle the inner city problem; by the early 1980s it had become associated more with the 'dead hand of local socialism' and was effectively bypassed by the new set of regeneration institutions epitomised by the UDCs; and by the early 1990s local authorities found themselves back in favour in so far as they were invited to act as the initial co-ordinators of City Challenge bids and latterly those under the Single Regeneration Budget (SRB).

It is clear that the antagonism which has characterised relations between central and local government over the last fifteen years has not made the task of local regeneration any easier. Both parties have given the impression that this battle has sometimes absorbed more of their creative energy and resources than local regeneration work itself. It is equally clear that local government is unlikely to lose its constitutional dependence on the centre for powers and resources and that the centre cannot afford to ignore local authorities completely in its attempts to do something about the inner cities. We must, therefore, remain hopeful that a more equitable and affable partnership characterises the relationship in the coming years.

The growth of new institutions

There has been an obvious growth in the number of new institutions charged with various aspects of urban regeneration in its broadest sense, to the extent that the Audit Commission felt obliged to comment on the complexity of the institutional environment, in the process developing the analogy of the 'patchwork quilt' (Audit Commission, 1991). Although widely quoted, this analogy is somewhat misleading as it implies that the various regeneration institutions were neatly stitched together, whereas the reality described by the Audit Commission was one in which there were significant gaps and overlaps more akin to a messy pile of blankets covering some parts of the body but leaving others exposed. So, in addition to the 'single-minded' Urban Development Corporations introduced by Michael Heseltine there have been Inner City Task Forces, City Action Teams and Safer Cities Units as well as local Training and Enterprise Councils and business leadership teams all trying to play meaningful roles in the stimulation and management of local economic growth (Bennett, Wicks and McCoshan, 1994; Robinson and Shaw, 1991). Stewart's chapter in this book (chapter 4) illustrates the changing patterns of expenditure associated with this proliferation of new institutions and in particular the switch of resources to them at the expense of local authorities.

While it is possible to see the emergence of many of these organisations as part of a concerted political attack on a relatively autonomous (and often Labour controlled) local government, the public justification for their introduction tended to be more pragmatic. By virtue of their size, composition, values and expertise, these new institutions were expected to be an improvement on the old and to deliver their policy goals more efficiently, economically and effectively. The verdict to date, in terms of both individual and collective impact, is one of mixed and muted success. To be sure they have been active, but it is not so clear that they have been successful in their own terms or more successful than their predecessors. While there can be benefits from institutional change and reorganisation, the pace and the frequency of these changes can be unhelpful to all concerned and a period of relative tranquillity would now be welcome (Lewis, 1992).

Measurement of success

As part of the more general trend of introducing principles, procedures and practices from business into the worlds of central and local government, there was a marked increase in the emphasis given to monitoring and evaluation in urban policy. The establishment of the Urban Programme Management Initiative in the mid-1980s heralded a renewed commitment to monitoring and evaluation, although its title, or at least the word 'management', indicates that its prime purpose lay more in managing locally administered grant regimes than in feeding local experience into national level policy development.

Since then there has been a view that urban policy monitoring and evaluation has focused too much on the development of intermediate output measures (such as the number of training places established or hectares of land improved) and too little on the impact or effectiveness of these activities in meeting policy goals (National Audit Office, 1990). Moreover, there is a suspicion that such intermediate output measures are of most use when presented in aggregate form in annual reports or in providing the stuff of ministerial soundbites. Having said this, recent reports by the Department of the Environment (1993) and the National Audit Office (NAO) (1993) have presented similar data on the performance of Urban Development Corporations which demonstrate a significant gap between targets and achievements and which have been used to call UDCs to account locally.

One of the more encouraging aspects of the City Challenge programme was the emphasis given to the development of a more comprehensive, ongoing regime of monitoring and evaluation. Preliminary analysis suggests, however, that the promises of bid documents have not always been maintained in subsequent action plans and annual programmes of work, especially in moving beyond the collection of routine output measures.

There are signs of a paradox here in the assessment of urban policy – measures designed to embody the pragmatic virtues of doing whatever is most effective have been able to ride out criticisms of their performance, even from 'respectable' critics such as the Audit Commission and the NAO, because of their political or ideological significance. We can only hope that the renewed welcome given to local authorities in playing a leading role in local regeneration is not part of a more cynical ploy of central government disengaging from responsibility for local ineffectiveness.

Competition for resources

Area-based inner city policies are sometimes seen as little more than traditional regional aid policies writ small, a smaller-scale combination of sticks and carrots for use with footloose capital. However, at the local level, competition for investment now clearly spans both the public and the private sector. City Challenge was the most transparent example of local authorities playing the game of maximising the promised return on investment. Not only do they now have to play this with potential private investors but also with central government investors in the Department of the Environment who are looking for the most plausible and secure returns on 'their' City Challenge money (Burton, 1991) and latterly on SRB monies.

It is probably still too early to start drawing sensible conclusions about the effectiveness of competition but we can say that it encourages both bold strategies and bold claims for the prospects of future success. In this climate it is likely to be increasingly difficult to assemble good quality data to assess the real performance of these new institutions and to use this to make sensible decisions about future policy directions. Many urban policy evaluators are now well practised in making few claims for any significant impact on the direction of policy. It is increasingly recognised that evaluation research can at best support developments already in train and there are few signs in recent years of political commitment to particular policy measures being shaken by any brute facts thrown up by research.

In conclusion, then, it is clear that urban policy over the last two decades has been characterised by a relatively high degree of consistency in its analysis of the problem and its basic approach to a response even though the policy rhetoric and institutional structures have changed substantially. Alongside this the social, economic and demographic context has seen a steady deterioration in the circumstances of those people targeted by urban policy – by and large the residents of inner city areas and peripheral housing estates. The failure of urban policy to achieve any significant improvement in the circumstances of these people is matched by a persistent belief among policy-makers that relatively small-scale, area-based regeneration measures involving local partnerships can succeed in transforming both the targeted areas and the people who live in them. Until there is a political acceptance of the need for much broader-based measures which link those disadvantaged in labour and housing markets back into the mainstream and which accept the civil right of all people to good quality public services, then we are unlikely to see such a transformation.

The amalgamation of virtually all urban policy measures into the new Single Regeneration Budget and the creation of newly integrated government offices in each of the standard regions to co-ordinate local programmes funded under it represents yet another major development in the institutional context of urban policy. The enhanced role of the new government offices and the creation of a new cabinet committee to decide which local programmes are accepted has been described as an injection of 'localism' into urban policy, whereby local people have the chance to develop their own programmes based on their own needs and priorities (DoE, 1994). The speed with which local proposals have had to be developed suggests that, like City Challenge before it, the new arrangements will see a greater commitment on paper than in practice to local involvement and devolved decision-making. The other notable feature of the new arrangements is the absence of any increase in the total resources available, resulting in a limited proportion of uncommitted money for new projects and programmes in the first year of operation. There are even suggestions that a second round of bids will not be invited (*Local Government Chronicle*, 1994).

Thus, not only does the context for urban policy seem to be unchanging, but the lessons of over twenty years of policy research and evaluation seem to have little impact as superficial changes to the administrative arrangements mask long-term continuities in a fundamentally flawed policy approach.

Acknowledgements

This chapter draws in part on work undertaken for the Department of the Environment in 1991–3 which examined socio-demographic and economic change in inner city areas immediately prior to the last Census (Boddy *et al.*, 1995). That work and this chapter represents the views of the authors and not those of the Department. Gary Bridge and David Gordon of Bristol University were also part of the original research team and their contribution is gratefully acknowledged.

3

AN EVALUATION OF URBAN POLICY

Michael Bradford and Brian Robson

Introduction

Beginning with the traditional Urban Programme in 1968, there has been a plethora of national government policies aimed at British inner cities and their problems. From the partnership and programme authorities under the enhanced Urban Programme in the late 1970s, through a series of new policy vehicles developed in the 1980s (Enterprise Zones, Urban Development Corporations, Task Forces and City Action Teams) to the more recent City Challenge (Atkinson and Moon, 1994), Single Regeneration Budgets, English Partnership and City Pride, national government has tried many mechanisms through which to deliver urban policy in an effective and efficient way. Over much the same period, governments have tried to cut or limit public expenditure and to demand greater value for the resources which they have invested. As an audit culture has begun to pervade the public sector and accountability has become a watchword, spending departments of government have felt it necessary to increase their monitoring and evaluation of such policy instruments. Although over the period many individual policy instruments have been evaluated (DoE, 1987a; DoE, 1987b; DoE, 1988; Ward *et al.*, 1989; DTI, 1991), there has been no overall assessment of the effects of urban policies.

The evaluation of individual policies has also been limited to direct outputs, such as the number of jobs created or the amount of private money attracted, and to efficiency, in terms of such measures as cost per job. There has been no assessment of the outcomes rather than the outputs of policy. By outcomes, we mean the much wider measures that reflect the quality of lives of those residing and working in the cities, such as reduced levels of crime and unemployment, and improved residential environments. Unlike direct outputs, outcomes are more difficult to measure and data for them are less available over both time and space. However, if we are to be able to judge whether urban policies are having a significant effect on the conditions of cities then an overall assessment of policy inputs on outcomes seems to be required.

The aims of this chapter are then to outline briefly the methodology of a large piece of research assessing the overall impact of urban policy, which was carried out for the government by a team of researchers in the Universities of Manchester and Liverpool, and to review some of its major findings (Robson *et al.*, 1994). First we discuss the conceptual problems that are tackled by the research design, and review the objectives of government programmes and the outcome indicators selected to represent them. The policy inputs are analysed and the relationship between these inputs and outcomes is discussed at a national and local level. Finally, reference is made to the results of a survey of residents and to in-depth interviews with local policy-makers and deliverers.

Research Design and Conceptual Problems

The brief given by the set of government departments on the project's steering committee was to evaluate the overall impact of inner city policies over the 1980s, paying particular attention to the packaging of policies in the late 1980s called 'Action for Cities' (AfC). The research was carried out on two major scales: the national, using a set of local authority districts; and the local, using the districts and wards of three northern conurbations, Greater Manchester, Merseyside and Tyne and Wear (Figure 3.1). The original proposal for the research design combined a mixture of extensive and intensive analyses which could inform one another. The quantitative work was aimed at establishing the strength of association between policy inputs and outcomes, while the qualitative research was directed at exploring the underlying processes and contexts, by and through which policy instruments worked with varying degrees of success or failure. The early quantitative work on three conurbations could suggest questions for in-depth interviews with policy-makers and -deliverers and with employers, and identify areas for questionnaire work with residents; while these interviews and surveys could in turn pose questions for the quantitative work on the relationship between policy inputs and outcomes. However, such a design would have taken two years and the eventual contract was for only eighteen months, so the quantitative and qualitative work on the three conurbations occurred simultaneously and subsequent to the quantitative research at the national scale (Figure 3.1). This national scale analysis of the 57 Urban Programme Areas (UPAs) was also prioritised by the government,

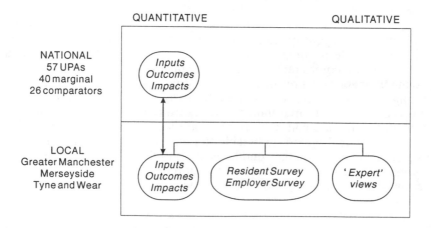

Figure 3.1   Design of the research.

because it provided greater areal coverage to inform policy, even if, without any qualititative analysis and less contextual understanding, it was bound to be of less depth.

There were numerous conceptual problems associated with this attempt to evaluate a multi-stranded programme of policies, which can be discussed under the heading of six 'Cs'. The most difficult question for any evaluation of policies is the *counterfactual* problem of assessing what would have happened in the absence of government intervention. A standard evaluative method would have been to compare the 57 UPAs where intervention has taken place with control areas that had similar problems but where there has not been any intervention. Such a method is not possible, because nearly all the districts with the greatest problems had been selected already for policy intervention. Instead, 40 'marginal' districts that were close to being included in the 57 were studied, along with 26 more 'comparator' districts which had similar, if not so extreme, problems. In total there were 123 districts for which the inputs, outcomes and impacts were compared (Figure 3.2).

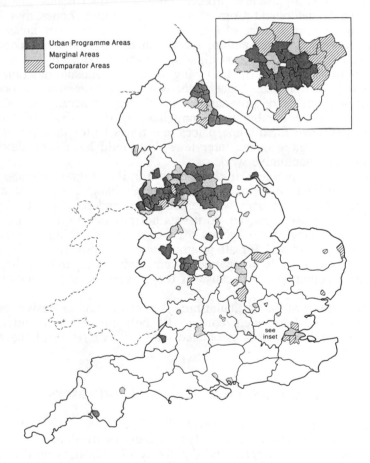

Figure 3.2   The sample 123 authorities.

The *confound* problem arises from the fact that outcomes can be affected by many public policies other than the specific government programmes included under the Action for Cities umbrella. The quantitative analysis tried to partition the effects of other programmes such as central government expenditure on mainstream programmes and European funds from the Action for Cities inputs as well as to account for their joint effects, while the intensive research also tried to account for the effect of the varying activity of local authorities.

The *contextual* problem concerns the very different conditions that local authorities were experiencing at the start of the period of study. These were likely to influence their capacity for improvement. In the quantitative analysis measures of these conditions were introduced as contextual variables, while in the intensive work the contexts of the three conurbations were thoroughly explored.

The *contiguity* problem in this study predominantly affected the local scale where intervention in one area can have either positive 'spillover' effects or negative 'shadow' effects on adjacent areas which are not subject to intervention. This is a particular problem with individual policy instruments such as Urban Development Corporations or Enterprise Zones, but is less obvious when studying a range of instruments because very few areas within inner cities receive no intervention. In the quantitative analysis contiguity effects could be identified by investigating residuals, while in the intensive work they could be explored by discussing areas with unusual outcomes.

The *combinatorial* problem concerns the way that intervention has occurred through different mixes of programmes in different places, and indeed at different times. Some combinations may have worked better than others, while some may have fitted some places and times better than others. This issue was explored more in the interviews, but could have been developed further within the quantitative work.

Finally there was a problem of *changes* both in the programmes and in the places targeted to receive them. The programmes changed over time as new ones were added and old ones were amended or ended. The set of places to which they refer also changed, so it was not possible to assign particular authorities to a 'policy-on' or 'policy-off' set in an unambiguous way over the whole time period and for all programmes. So although the 57 UPAs, the 40 marginals and the 26 comparators were analysed separately, they were also analysed together, partly because of these overlaps in designations and programmes.

So both our choice of the particular mix of intensive and extensive research and the form of the latter were consciously determined by our attempt to tackle these difficult problems associated with the evaluation of the effects of policy.

Programmes, Objectives and Indicators

Although the research design has been outlined, we have not yet stated how we set about the evaluation process. This is most easily done in discussing the programmes, their objectives and the choice of indicators used to reflect them. There were many policies listed under the Action for Cities initiative

(Table 3.1). Sometimes it was not always clear whether a particular policy was regarded as part of inner city policy or not; for example, the Department of Trade and Industry (DTI)'s Regional Selective Assistance has its origins in regional rather than inner city policy. A preliminary investigation of these policies (Robson *et al.*, 1991, summarised in Robson *et al.*, 1994) revealed that most were targeted spatially rather than at specific social groups. It was this that prompted the research design. The spatial targeting, however, occurred at varying scales and was directed at different spatial units even at the same scale.

Table 3.1   Programmes included in determining the objectives of policy

1. Urban Programme
2. Urban Development Corporations
3. Enterprise Zones
4. City Grant
5. City Action Teams
6. Derelict Land Grant
7. Housing Corporation
8. Land Register
9. Housing Action Trusts
10. Estate Action
11. Garden Festivals
12. Safer Cities
13. Section 11 Grants
14. Ethnic Minority Business Initiative
15. City Technology Colleges
16. (Inner City) Open Learning Centres
17. English Estates' Managed Workshop Programme
18. Task Forces
19. Enterprise Initiative
20. Regional Selective Assistance
21. Transport Supplementary Grant
22. Jobclubs
23. Loan Guarantee Scheme
24. Race Relations Employment Advisory Service
25. Small Firms Service
26. Employment Training
27. Enterprise Allowance Scheme
28. (School-Industry) Compacts
29. Youth Training Scheme
30. Headstart

The notion of 'inner cities' clearly varies both within and between different programmes: the Urban Programme was directed either to whole authorities or to parts of authorities among the 57 UPAs; Enterprise Zones (EZs) may or may not have been targeted to UPAs; some programmes such as Estate Action could apply to any authority even though in practice there was a focus on UPAs. Such uncertainty is a function partly of when policy instruments came on stream, partly of an ambiguity in the very concept of 'inner cities' and partly of the objectives at which policy was aimed.

The documents associated with the programmes were all examined to

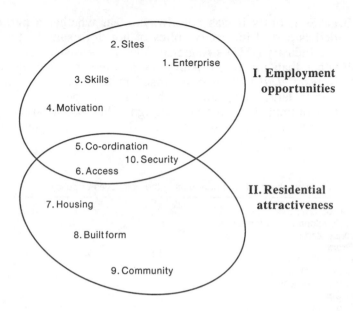

Figure 3.3   Higher-level and lower-level policy objectives.

discover their objectives. These were mostly implicitly rather than explicitly stated. Some 100 objectives were uncovered and these were classified under ten headings and mapped in a Venn diagram to yield two major overall objectives: to improve employment opportunities and to enhance residential attractiveness (Figure 3.3). These objectives informed the qualitative work and also formed the basis of the choice of outcome indicators on which we could measure the overall impact of policy inputs in the quantitative section. The choice of indicators is affected by their availability over both time and space. They had to be available at a sufficient number of intervals during the 1980s to make some meaningful comparisons of places before a particular round of policies, or on either side of a peak or trough in national economic conditions. The indicators chosen are not without considerable disadvantages (Robson *et al.*, 1994), but they do give a reasonable coverage of the identified objectives and they are available for a number of points of time for most places, and in some instances at different spatial scales. The selected indicators are: unemployment, long-term unemployment and the ratio between the two, for 1983 to 1991; net job change from 1981 to 1989 (intermediate points 1984 and 1987); change in small businesses using value added tax (VAT) registrations, 1979–90; house price change for first-time buyers and inner area property for 1983 to 1990; and net change in the number and proportion of 25–34-year-olds for 1981–90. The first three sets of indicators reflect the employment opportunities objective, while the last two both reflect this indirectly and measure the attractiveness of the areas as places to live. The last indicator selects out the 25–34 age group, because it can be argued to be a particularly significant one in affecting and reflecting the economic and social health of the city. Their in-migration to and retention in inner cities is considered to be an important sign of improvement, unless it

Table 3.2 Expenditure in the sample of 123 districts: AfC and 'total' expenditure

	1979/80	80/1	81/2	82/3	83/4	84/5	85/6	86/7	87/8	88/9	89/90	90/91
AfC	*156*	*211*	*283*	*402*	*462*	*488*	*591*	*672*	*812*	*1115*	*1321*	*1099**
(of which)												
UP	120	163	170	221	222	232	226	224	242	227	230	233
UDC		30	67	94	91	93	109	168	316	445	588	
EZ			32	26	63	115	111	96	171	207	NA*	NA*
UDG/CG				1	31	18	36	33	48	60	64	52
CATs								2	5	7	9	12
DLG	10	15	31	34	32	42	47	51	43	41	43	
EA									127	170	NA*	NA*
SC										2	5	
S11	26	42	52	60	55	59	67	68	77	80	83	
EMBI							0.2	0.1	0.2	0.2	0.2	0.2
EE							26	26	37	42	36	
TF							12	30	45	35	41	41
HIP	1,628	1,449	1,153	1,451	1,414	1,221	1,056	1,034	996	988	827	1,309
RSA/RDGII	39	40	45	51	44	40	62	100	138	228	165	131
ERDF/ESF				265	178	157	131	137	109	33	35	
RSG	4,225	4,963	5,297	5,383	5,416	5,639	5,578	5,884	6,199	6,304	6,438	
TOTALS	**6,047**	**6,662**	**6,779**	**7,244**	**7,569**	**7,646**	**7,505**	**7,514**	**7,967**	**8,639**	**8,650**	**9,011**

Notes:

1. All figures are cash values and are shown in £ millions.

2. *NA. When calculations were made, disaggregated data were not available for Enterprise Zones or Estate Action for 1990/91. The AfC total for that year is therefore necessarily artificially depressed.

3. The total European funding cannot be disaggregated for districts for the whole period: ERDF runs from 1983/4 to 1988/9; ESF runs from 1987/8 to 1990/1.

4. Abbreviations:

AfC: Action for Cities (of which)
UP: Urban Programme
UDC: Urban Development Corporation
EZ: Enterprise Zone
UDG/CG: Urban Development Grant, Urban Renewal Grant, City Grant
CATs: City Action Teams
DLG: Derelict Land Grant
EA: Estate Action

SC: Safer Cities
S11: Section 11, Education
EMBI: Ethnic Minorities Business Initiative
EE: English Estates
TF: Task Force

HIP: Housing Investment Programme
RSA/RSGII: Regional Selective Assistance/Regional Development Grant II
ERDF/ESF: European Regional Development Grant/Social Fund
RSG: Rate Support Grant

is accompanied by a proportionate or greater-than-proportionate increase in the long-term unemployment of the age group which would suggest that they are being trapped there rather than being attracted in or to remain. These indicators are coarse and there are major caveats attached to them, but they do reflect the objectives and outcomes.

The evaluation process then entails: first, an assessment of the extent to which government has achieved its own implicit objectives by its policies, using both the quantitative analysis comparing the relationships between inputs and outcome indicators and the in-depth interviews of deliverers; and, secondly, an assessment of the extent to which key people, employers and residents think that inner cities are improving/declining relative to other areas, and their views of the success/failure of policies according to their own criteria. This is therefore a standard evaluation of the degree to which stated objectives have been achieved, together with a broader perspective from those involved with or those receiving policies, and of course our synthesis of the two.

Policy Inputs

We have already indicated that programmes came on stream at different times (Table 3.2). We were able to discover from various government departments the amounts of money involved in every programme each year and the places to which it was allocated. Our detailed analysis only includes that part of inner city policy expenditure that can be assigned to districts. The training programmes, for example, are omitted because they could not be allocated below the regional level. The total amount spent was quite considerable and increases over the decade (Table 3.3). Although inner city expenditure is quite large and some districts received considerable sums, many also faced major reductions in their mainstream finance; for example in housing monies through the Housing Investment Programme (HIPs) and their general recurrent allocations, the Revenue Support Grant (RSG). In many cases this overall reduction was much greater than the total amount of inner city expenditure. It would therefore not be surprising if for some places inner city policy inputs had little effect on outcomes.

The fact that as part of our research we had to assemble and aggregate these inputs in itself suggests that national government had not monitored in this overall way where its monies had been going. This may not be surprising given the constraints of time, but it does not bode well for the degree to which the objective of spatial targeting of the overall policy is met.

Table 3.4 shows the rankings of the top 79 of the 123 districts according to the total amount of expenditure on Action for Cities programmes between 1988/9 and 1989/90. A similar message would be conveyed by data on the whole of the 1980s. Of the top 13 recipients, 4 were not included within the 57 UPAs. At the other extreme 10 of the 57 UPAs did not rank within the top 57 districts for expenditure. Indeed it is the 79th rank before the last of the 57 appears. So the targeted areas were not necessarily receiving the most money.

The same picture emerges at the local level when examining the total amounts allocated to various wards. Some priority areas were receiving less monies than non-priority ones. So without any monitoring the supposed targeting is less effective than it might be.

The Relationship Between Inputs and Outcomes
National scale

The quantitative modelling explored the degree of association between inputs and outcomes, allowing for the contextual variables describing the initial

Table 3.3 Officially recognised overall expenditure on Action for Cities and expenditure included in the CUPS analysis

(£ Million)	1988/89	1989/90	1990/91
* Urban Programme	261	260	261
* Urban Development Corporations	234	439	542
* City Grant	25	35	49
* City Action Teams	7	6	8
* Derelict Land Grant	31	20	21
* Estate Action	140	190	190
* Section 11	57	69	73
* Task Forces	16	19	23
* Safer Cities	0.2	2.5	6.5
* Ethnic Minorities Business Initiative	-	-	-
* Enterprise Zones	-	-	-
* English Estates			
***Regional Selective Assistance	189	188	192
** Housing Investment Programme(ICSA)	-	-	100
***Housing Corporation	610	517	620
Housing Action Trusts	-	-	42
Training/Enterprise	1,096	1,035	1,040
Homelessness	-	-	147
Transport Supplementary Grant	190	220	300
Community Technical Colleges	14	32	45
Scotland and Wales	300	350	400
TOTALS			
Total government expenditure	**3,170**	**3,382**	**4,060**
CUPS AfC evaluation expenditure (For details, see Table 3.1)	**1,115**	**1,321**	**1,099**

Notes:

1988/89 data are out-turn; 1989/90 data are estimated out-turn; 1990/91 data are planned figures.
* Programmes included in CUPS's evaluation research definition of 'AfC expenditure'. (Official figures do not include Enterprise Zones or EMBI expenditure in the AfC expenditure and do not show expenditure on English Estates separately.)
** Programmes included in CUPS's evaluation research definition of 'overall expenditure'.
*** Programmes for which output rather than expenditure data are included in CUPS's evaluation research (see Appendix A of Robson *et al.*, 1994).
# Spatially disaggregated data for Enterprise Zones and Estate Action for 1990/91 were not available at the time of calculating input/outcome relationships. This accounts for the lower figure for the CUPS totals for 1990/91.

Table 3.4 Targeting of AfC expenditure: ranking of the top 79 of the sample of 123 districts by per capita AfC expenditure, 1988/9 and 1989/90

Rank District	Per capita	Designation (with main expenditure resource head for non-UPAs within the top 57)
1. Tower Hamlets	1,717.11	UPA
2. Newham	799.51	UPA
3. Scunthorpe	269.27	non-UPA (EZ)
4. Southwark	225.56	UPA
5. Gateshead	206.07	UPA
6. Corby	189.35	non-UPA (EZ and LG)
7. Salford	183.26	UPA
8. Stockton-on-Tees	177.02	UPA
9. Hartlepool	170.82	UPA
10. Newcastle-on-Tyne	148.63	UPA
11. Trafford	134.31	non-UPA (EZ and DC)
12. Sandwell	132.40	UPA
13. Rochester-on-Medway	121.47	non-UPA (EZ and E)
14. Middlesbrough	117.64	UPA
15. Liverpool	115.95	UPA
16. The Wrekin	105.86	UPA
17. Manchester	102.44	UPA
18. Dudley	98.58	UPA
19. Kingston-on-Hull	98.06	UPA
20. Walsall	96.87	UPA
21. Islington	89.53	UPA
22. South Tyneside	89.10	UPA
23. Rochdale	87.38	UPA
24. Blackburn	86.18	UPA
25. Rossendale	85.86	non-UPA (EZ,DLG,EA and EE)
26. Wolverhampton	79.88	UPA
27. Burnley	79.75	UPA
28. Hyndburn	77.67	non-UPA (EZ,CG,DLG and EA)
29. Preston	71.23	UPA
30. Sunderland	70.83	UPA
31. Oldham	68.96	UPA
32. Coventry	68.44	UPA
33. Nottingham	65.65	UPA
34. Knowsley	65.54	UPA
35. Birmingham	61.10	UPA
36. St Helens	60.12	UPA
37. Sheffield	58.87	UPA
38. Wirral	57.66	UPA
39. Hackney	54.76	UPA
40. Leicester	52.82	UPA
41. Bradford	51.94	UPA
42. Rotherham	51.69	UPA
43. Lambeth	50.61	UPA
44. Halton	49.43	UPA
45. Haringey	49.06	UPA
46. Kensington/Chelsea	46.62	UPA
47. Greenwich	46.58	UPA
48. Bolton	43.58	UPA
49. Brent	43.58	UPA
50. Derwentside	43.40	non-UPA (DLG and EE)
51. Barnsley	42.80	UPA
52. Gravesham	42.69	non-UPA (EZ)
53. North Tyneside	41.85	UPA

54. Ellesmere	41.31	non-UPA (DLG and EE)
55. Doncaster	41.17	UPA
56. Wellingborough	39.64	non-UPA (EZ)
57. Langbaurgh	38.94	UPA
58. Derby	38.28	UPA
59. Barrow	34.60	
60. Wigan	33.89	UPA
61. Pendle	33.63	
62. Leeds	32.36	UPA
63. Hammersmith/Fulham	32.30	UPA
64. Ealing	27.89	
65. West Lancashire	26.39	
66. Wandsworth	25.23	UPA
67. Kirklees	24.53	UPA
68. Lewisham	23.33	UPA
69. Bristol	21.98	UPA
70. Wakefield	20.98	
71. Waltham Forest	20.92	
72. Easington	17.76	
73. Chester-le-Street	16.51	
74. Calderdale	15.45	
75. Stoke-on-Trent	12.05	
76. Sefton	11.00	UPA
77. Enfield	10.59	
78. Bury	10.40	
79. Plymouth	10.15	UPA

Note: Details are shown for the top 79 authorities so as to include all of the 57 UPAs.

conditions within the districts. Any interpretation of the results of such an investigation are not unambiguous. If the association were negative, it could mean that the inputs were ineffective or it could be that the inputs were correctly targeted, but that the response to an economic upturn was at a slower rate than elsewhere, but would have been even slower without the intervention. If it were a positive association, public assistance could be seen to have had beneficial effects or it could be said that the resources were inappropriately targeted and that the places might have recovered without the intervention. Such quantitative analysis only establishes the association. It is the intensive research that is more likely to reveal which interpretation is more appropriate.

Various forms of analysis were employed, from highly sophisticated multi-level modelling, through traditional regression, to more mundane analyses of the changing gaps between the 57 and the rest of the 123 and between inner city wards and the rest of the conurbation. Only the findings and not the techniques are reviewed here. (For the techniques, see Robson *et al.*, 1994.)

There was some evidence of a positive association between certain inputs, namely Urban Programme monies and Regional Selective Assistance, and improvement in the outcome indicators, especially the unemployment ones. This was the case between 1986 and 1990, a period when unemployment within the 123 as a whole was declining. During the previous period of increasing unemployment in the 123, 1983–86, there had been little relationship. In all cases the tests were carried out with a lagged relationship between inputs and changes in outcomes; for example, the inputs of 1984/5 to 1988/9 for the 1986–90 changes. Since change in unemployment after 1986 was related to the 1986 levels, the tests of the relationships between inputs and outputs from 1986–90 also allowed for these initial 1986 levels, and the relationship between inputs and outcomes was still statistically significant.

A less sophisticated analysis compared the average rates of the indicators for the UPAs at the start and end of the period with those for the marginals and comparators. This analysis was carried out for the North West and Greater London, where a number of marginals and comparators had been included (Figure 3.2). Although for unemployment it would have been better to compare the places at similar points of the economic cycle, data availability only allowed a comparison of 1983 with 1991, both low points but not necessarily the lowest points of the cycle for these places. For both the North West and Greater London, two very different regions in terms of unemployment changes, the average gap between the UPAs and the others had closed. Since the amount of inputs had been greater in the UPAs, it could be argued that again there were signs that intervention was having some success.

However, the variation among the UPAs both in changes of outcomes and amounts of inputs suggested that an analysis of average gaps concealed much. Further analysis examined the individual UPAs and classified them according to their general changes in outcomes and their inputs. The inputs included not only those directly related to inner city policies but also the important changes in the mainstream programmes of HIP and RSG.

Figure 3.4 summarises the position of the 57 districts and some important marginals (m) on these dimensions. Although generally most of the districts occupy the top left to bottom right diagonal, again suggesting the important role of intervention, there are also a significant number of outliers from that trend. In the top right corner there are the districts into which considerable resource has been allocated, but with little obvious improvement in outcomes. These include districts in the conurbations, such as Liverpool, Knowsley, Newcastle and Sunderland as well as some London boroughs and large cities. These remain a continuing problem, but of course might have experienced even greater problems in the absence of resources. Although there are no equivalent occurrences in the bottom left of the diagram, there are some districts where relatively small amounts of inputs have had significant benefits, at least according to the outcome indicators used in this analysis. Returning the diagonal trend, the bottom right location shows those districts which, although targeted, have received relatively little net input of resource and have shown relatively little improvement. They include the central districts of conurbations, such as Birmingham, Manchester and Sheffield, major cities such as Leicester and Nottingham and some London boroughs. Clearly these need to have a much larger net increase in inputs in order to have much chance of improvement. On the other hand, the top left location shows the successes of government policy. Corby with its enterprise zone is not an UPA, but it has made dramatic improvements in the outcome indicators. So too has Tower Hamlets, the district with the greatest amount of input over the period. Although it shows up well on this district level analysis, other work and more local analysis suggests that this 'success' conceals some very poor pockets of socially disadvantaged groups (Bradford et al., 1993).

The final point to be made about this matrix is that the regional location of the district has little relevance to its position on the table. Newham and Tower

Hamlets are located next to one another and, as parts of the London Dockland Corporation, received very large amounts of government resources. Yet at the district scale, one shows considerable success while the other remains a major problem. Similarly districts in the North East occur in different parts of the figure: Newcastle in the top right; Hartlepool and Stockton top left; North Tyneside towards the bottom left; and South Tyneside and Langbaurgh in the bottom right. The regional context, therefore, does not seem in itself to be a determinant of the degree of success of urban policy.

This specific comment on spatial scale highlights the more general concern about the scale of the delivery and scale of the analysis of urban policy: the district masks so much internal variation both in degree of problems and degree of improvement. Inner-conurbation districts such as Leeds which may be said to be over-bounded, because they include outer suburbs and indeed semi-rural areas, tend to rank less poorly on measures of deprivation than more closely bounded districts such as Liverpool and Manchester, yet they have very intense pockets of deprivation (Bradford *et al.*, 1993). Equally such districts may seem

OUTCOMES

INPUTS	POSITIVE	REL. HIGH	MIXED	REL. POOR	POOR
HIGH	Corby (m) Tower Hamlets				Newham
REL. HIGH		Dudley Gateshead Greenwich Hartlepool Sandwell Scunthorpe (m) Stockton Trafford (m) Wandsworth Wrekin			Coventry Hackney Hull Knowsley Lambeth Liverpool Newcastle Sunderland
MIXED		Blackburn Bradford Derwentside (m) Hyndburn (m)	Brent Ellesmere (m) Hammersmith Preston Rochdale Rochester (m) Rossendale (m) Salford Southwark St.Helens Wirral	Haringey Leeds	
REL. LOW		Bristol Burnley Middlesbrough North Tyneside Pendle (m) Sefton Walsall Wellingborough (m) Wolverhampton	Bolton Derby Doncaster Halton Kensington Kirklees Oldham Plymouth Rotherham Sedgefield (m) Tameside (m) Wigan		
LOW					Barnsley Birmingham Blyth (m) Camden (m) Islington Langbaurgh Leicester Lewisham Nottingham Manchester Sheffield South Tyneside

Figure 3.4 Inputs: outputs – a classification of districts.

to respond well to intervention and yet it may be the better-off areas within them that are improving and not the pockets of deprivation. It was for these reasons, among others, that the three conurbations were examined at a finer spatial scale.

Local scale

The analysis of the conurbations included both quantitative and qualitative analysis. First at the scale of the conurbations it is worth noting that, over time, unemployment, long-term unemployment and job loss were increasingly concentrated in the central districts of Liverpool, Manchester and Newcastle. They had an increasing share of these relative to their conurbations. So the relative position of these central districts in their conurbations seems to have been worsening, even though there was some absolute improvement in their position and that of the conurbations as a whole.

Within districts we analysed wards according to their location within or outside the inner cities. There was an increasing polarisation of unemployment; inner areas in 17 of the 20 districts increased their share of unemployment over the period. This was not the case for Manchester and Liverpool, where if anything, one could regard the inner cities as having expanded in area. As an indication of the worsening relative position of the inner city wards, in 19 out of 20 districts their inner city wards had a greater increasing share of the districts' long-term unemployment than of unemployment.

These polarisation effects suggest a bleaker interpretation of the national analysis at the district scale. They suggest that in at least some districts the improvement is largely in their better-off areas rather than inner city wards. This would be true, for example, of the district results for Tower Hamlets.

These findings also suggest that the major underlying process of inner city policy throughout the 1980s, namely 'trickle down' is not working. The major concept, imported from the USA, was of property-led development in general and flagship developments in particular which would attract private investment. Although these developments would often not directly affect local disadvantaged people, with for example jobs going to better-off commuters, the argument was that they would have beneficial effects indirectly through trickle down. Such a process is rarely detailed, but whatever forms it might take, it does not seem to be working in the English inner cities.

Second, we looked at the input/outcome relationship at a sub-district scale by assembling expenditure data for wards and examining the relationship between these inputs and indicators of social change. The allocation of expenditure to wards was necessarily imprecise because much of the resource was spent on district-wide expenditure. Nevertheless, the database on which we draw probably represents the best possible estimate of sub-district expenditure patterns.

The relationship between the inputs and outcomes at a ward level was analysed in a sophisticated way. The outcome indicators were unemployment change between 1983 and 1986, and between 1986 and 1990, and job change per capita between 1984 and 1990. When wards were grouped according to their inner city characteristics at the start of the period, there was too much variability within groups in their outcome indicators to allow the analysis to continue to account for difference between groups by policy intervention. Indeed the above analysis of outcomes showed great variation in the experience of individual wards within the inner cities. So whether wards belonged to inner city areas or not does not give any significant explanation of changes in outcome indicators. Even when wards were grouped according to their changes in outcome indicators, regardless of their location within the district, there was still no significant relationship between the inputs and the outcomes. In short, there was little statistical evidence from this analysis of the success of policy inputs at this scale.

Survey and Interviews

Surveys of residents were also carried out in particular areas of the three conurbations. Areas were chosen in pairs, one of which had received a considerable amount of government input and a neighbouring area which had either received less or was felt by local experts to have had less successful policy intervention. This design was selected to see whether the experience and expectations of residents in the two adjacent areas were different. The aims of the interviews were to gauge the residents' views on some changes that were difficult to measure in the quantitative analysis, namely safety from crime, the visual appearance of the residential area and more generally its overall attractiveness as a place to live, together with changes in employment opportunities, the second major government objective, which had been included in the quantitative analysis.

The major findings of this survey of 1,300 residents were that: areas where there had been more policy inputs were viewed in more positive ways by their residents; thus from the residents' viewpoints the policy inputs had been worthwhile. Despite this, the general view of the past few years was of decline and a widening gap with the outer suburbs, while the view of the future was pessimistic, especially about levels of crime and employment opportunities.

To give some examples, first, two areas of rather different amounts of input are discussed. More residents of Ordsall in Salford rated their area as desirable than residents in Pendleton and many more thought that it had improved in the last three years. Even more significantly, more residents in Pendleton thought that Ordsall would improve in the future than thought that their own area would improve, while even more residents of Ordsall saw a positive future for their own area and even fewer for that of Pendleton. Similar results occur for Dingle and Granby in Liverpool. Here both areas have received considerable amounts of inputs, but local experts thought that policies had been more successful in Dingle than in Granby. The views of the residents fully justified the experts' analysis. Finally the more recent initiatives in Cruddas Park (West City) in Newcastle and Town End Farm in Sunderland seem to have had a greater impact than the longer period of initiatives in Scotswood and Southwick, their respective members of the pairs. In Cruddas Park, for example, there has been significant recent expenditure under the Urban Programme, Estate Action and City Challenge, with about three times the amount being spent there between 1988 and 1991 than in Scotswood, which had received much more money from the Urban Programme in the early to mid-1980s.

Some findings from these surveys are echoed in the in-depth interviews of local policy-makers and deliverers. The first concerns the local residents themselves as a resource. Many of those interviewed thought that too many policies had ignored the local residents or more generally the local community. The surveys showed that despite the general feeling of decline and pessimism, there was considerable loyalty to their local area which emerged in their comments about its residential desirability, especially compared to a neighbouring area. It would be all too easy to reflect a 'grass is greener view', but quite the reverse was true. This 'place loyalty' can be built upon and the local community therefore suggests itself as a much ignored local resource.

Secondly the interviews suggested that where local people had been more involved in the policy inputs they had proved more successful. This was supported by the evidence of the survey in Cruddas Park in Newcastle and Miles Platting in Manchester, where small-scale, locally tailored initiatives are beginning to tackle the deeply ingrained problem of long-term male unemployment by building confidence and 'getting people into jobs' (Hayton, 1990). More generally ways need to be found to empower people. The emphasis of the 1980s on economic and environmental intervention has not sought to tackle social disadvantage in a sufficiently direct way.

The interviews also suggested that these 'local blackspots' of poverty had become cemented to an even greater degree than in earlier decades. These pockets of disadvantage and the wider structural economic issues that affect major cities producing, as one local policy-maker commented, a near permanent base-load unemployment rate of around 1 in 10, are thought to require much greater intervention. 1980s' policies were felt to be no more than 'sticking plaster' which could only have marginal impacts relative to the deep and intractable problems of conurbations.

Further evidence from the interviews suggests the need for a stronger role for local government, a view expressed not only (unsurprisingly) by experts in local authorities, but also by many of the experts from other sectors. There is sufficient evidence from non-local government people to suggest that greater local understanding and more locally based initiatives could contribute much to regeneration.

Certainly there is evidence from the interviews of major differences in local political contexts and attitudes, and the effects these have on the success of policies. For example, it is interesting to note the differences among the conurbations in their views about the public sector in general. In Tyne and Wear, with its traditional right-wing Labourism and the interventionist bent of many of the region's business community, there is a stronger consensus that the public sector should play a lead role in urban regeneration. Here there was much stronger support for traditional and possibly local forms of Keynesian intervention. The limitations of property-led regeneration through such instruments as Urban Development Corporations and English Partnership in times of a recession in the property market were often reported and other forms of regeneration were preferred. Some wanted to spend through a recession. Others wanted public sector intervention to remove some of the uncertainties for the private sector.

It was here in Tyne and Wear where there seemed to be a much more coherent set of local government officers and councillors, private business people and people from the voluntary sector that were playing an important role in the regeneration of the conurbation. In the other two conurbations there was much less evidence of a coherent approach. This perhaps accounts for the shared hesitation in Tyne and Wear to a more formal regional institutional structure, which indeed has emerged since our report was submitted. There was already a sense of 'collective togetherness', which gave some structure to central government inputs. They considered that they already had a successful existing local network of institutions which co-ordinated and delivered policy. National government's response to improve regional co-ordination, perhaps, has more to do with other regions which are less well organised, and with co-ordinating its own programmes more effectively, a common criticism among interviewees.

There was some evidence of successful implementation of policies. The idea of a multi-agency approach, for example, was seen to have explained the success of the Tyneside Enterprise Zone and seen as a major advantage of City Challenge. Yet others suggested that the particular policy arena of the North East with the active 'hands on' approach to management of the Enterprise Zone adopted by English Estates accounted for the particular success of this, as against other enterprise zones in the country. Again the political context of the conurbation seems an important element in the success of policy.

Conclusions

The various arms of the research suggest some limited success for government policy. The quantitative analysis reveals an association between inputs and improvements in outcomes at the national level for the late 1980s. The gap analysis suggests some closure between UPAs and the rest of the 123 districts. Smaller cities and the outer districts of conurbations seem to be more successful than larger cities and districts at the centre of the conurbations. At the local level the residents survey indicates the relative success of some policy inputs for small areas within 'inner cities'. Finally the interviews suggest that some policy instruments have worked well, particularly when there has been a multi-agency approach without inter-agency conflicts, and where the local policy arena has been active and well co-ordinated.

On the other hand, the amount of money going into urban policy is miniscule compared to the size of the problems which are being tackled, and the loss of mainstream money in many authorities has more than countered any increase in urban funds. This loss has also limited the use of urban funds and severely constrained local initiatives. Although government inputs were supposed to be spatially targeted, the lack of monitoring in part has led to rather inefficient targeting at both the national and local level. The question of the consistency of funding for areas with deeply ingrained problems is further raised by more recent initiatives, such as City Challenge.

Although some of the 57 districts show signs of improvement as a whole, there is considerable other evidence to suggest that many of the poorer areas are not improving or at least not nearly as much as the better-off areas within the districts. The property-led developments have not produced any trickling down effects for these areas. The economic and environmental emphases of policies have not tackled these areas of social disadvantage, where the place loyalty of the local residents suggests that there is plenty of human capital in which to invest effectively.

The capacity of the local community is not the only resource that has been under-used. Both local government and the local voluntary sector can play much greater parts in future regeneration and can attempt to produce effective local policy arenas into which central government policy and the activities of its new regional offices can integrate.

Finally, although there are generalised findings made in the report, it must be re-emphasised that there is much variation in the success of overall policy within England, within regions, within conurbations and even within the wards of an inner city. The task of unravelling precisely why there is such great variation is only partly answered by this report. Just as there is much still to be done in regenerating cities, so too there is much research still to be carried out on why sets of policies are more successful in some areas than in others.

Acknowledgements

We would like to thank Michael Parkinson, Ian Deas, Ed Hall, Eric Harrison and Peter Garside for their assistance with this research project, the Department of the Environment for their funding, and Graham Bowden and Nick Scarle of the Cartographic Unit in the School of Geography, University of Manchester for their work on the maps and diagrams.

4

PUBLIC EXPENDITURE MANAGEMENT IN URBAN REGENERATION

Murray Stewart

Introduction

Urban policy is usually evaluated against the substantive objectives to which the instruments of policy are addressed – employment creation, provision of housing, renewal of the built environment and so on. Financial resources are an input to, rather than output from, such policy intervention. Nevertheless the planning, management and above all control of public expenditure has been a key feature of government policy over the past twenty years. Indeed the containment of public spending has of itself been a specific and central objective of public policy over the period, not only of the successive Conservative regimes of the 1980s and early 1990s but also of the preceding Labour administration.

Public expenditure management is the focus of this chapter. The first step is a mapping of the shifts in actual and planned urban policy public expenditure over the period 1988–89 to 1996–97. This is essentially a descriptive exercise drawing upon the Government's public expenditure planning documents and most notably the annual Department of the Environment Reports on spending plans which for four years have replaced the Public Expenditure White Paper (DoE, 1991; DoE, 1992; DoE, 1993; DoE, 1994). A second section of the chapter examines the transition from separate programme budgets to the Single Regeneration Budget and considers the implications of this transition for the future management of regeneration policy. One theme is the reduction in public resources and the consequent implications of scarce resources for the leverage of private sector funding and finance. The third part of the chapter, therefore, examines the history of those elements of the urban spending block which in the recent past have involved interdependence with the private sector – Urban Development Corporations, City and Derelict Land Grant regimes, the New Towns Programme – and discusses both the transparency and predictability of public expenditure plans and outturns. A final section draws conclusions from the analysis.

A number of factors – not least the emergence of urban violence in the early to mid-1980s – combined to partially protect 'urban' expenditure from the most severe expenditure cutbacks. Indeed from 1981–82 to 1991–92 Urban

Programme expenditure increased by 40 per cent, whilst net expenditure on Urban Development Corporations rose to over £600 million. By 1992–93 the cash total of all of what were then called the 'inner city' programmes amounted to over £1 billion. Similarly official estimates of the levels of historic spending within what is now to be the Single Regeneration Budget indicate that from 1981–82 to 1993–94 expenditure within the twenty relevant programmes rose (in constant prices) from £414 million to the 1994–95 planned level of £1,443 million.

Total central government financial support as a whole for cities is a different question, of course, with the main programmes, the bending of which had been such a strong feature of the 1977 White Paper on Inner Cities, suffering severe reductions.

> Most local authorities experienced real or relative reductions in both Housing Investment Programme and Revenue Support Grant finance during the decade . . . Such reductions meant that, even though Action for Cities resources may have increased or held steady (and thereby became an increasingly important part of total resources), such authorities experienced significant overall reductions in the public resources available to them.
>
> (Robson *et al.*, 1994)

Nevertheless the latest PSI Urban Trends report (Policy Studies Institute, 1994) argues that Revenue Support Grant (formerly Rate Support Grant), measured in terms of per capita grant, increased between 1981–82 and 1994–95 in thirty deprived boroughs/districts. Indeed the increases in these areas were larger than the average increase for all English local authorities. Whilst in 1980–81 grant per head in the deprived areas was below the English average, by 1993–94 the ratio had shifted to the point where the grant per head for the thirty deprived areas exceeded the English average by 22 per cent. Per head averages of course conceal the differing needs of different areas as assessed either by central government or by localities themselves. Per capita figures also fail to capture the impact of falling population levels upon the level of resources available to local authorities or on the levels of spending need. Nevertheless it would be wrong to characterise the 1980s exclusively as a period when resources were curtailed – certainly as far as the formal urban programmes of the Department of the Environment were concerned.

As far as urban policy is concerned the 1980s were characterised as much by budgetary invention as by fiscal austerity. Two main features dominate the decade. On the one hand there was a fragmentation of traditional institutional form with the proliferation of new agencies and organisations holding a variety of responsibilities for urban intervention. At the same time there was throughout the 1980s an increasing reliance upon private sector partnership, and most explicitly upon development interests as the engine of urban regeneration. These mutually reinforcing factors – the containment of public expenditure together with institutional fragmentation and collaborative dependency upon the private finance and development industry – combined to transform the public expenditure allocation and control processes as they affected urban programmes.

In 1980 there remained a relatively stable programming mechanism still characterised by partnership, programme and 'traditional' urban programme spending. The central/local relationship was one of annual bidding for resources within a well understood if burdensome administrative system (Stewart, 1987). The Urban Programme at that time was subject to considerable criticism over the lack of clear monitoring arrangements as well as over the absence of a strategic framework within which the very large number of individual projects were initiated and implemented. In 1991–92 there remained almost a thousand separate projects supported by the Urban Programme (Hansard, 1994). The value for money issues in the early years of the 1980s were thus whether DoE actually knew exactly where expenditure was going. Administrative innovation was directed at the development of the Urban Programme Monitoring and Information (UPMI) system.

Over the next ten years the emphasis in public expenditure planning and control shifted towards a resource allocation regime which in practice supported the fragmentation referred to above and which was thus directed to the funding of new initiatives and to the allocation of resources between a range of at best complementary and at worst competing institutional forms and policy instruments. Whilst successive initiatives aspired to integration – Task Forces, City Action Teams and latterly City Challenge – in practice each initiative acquired a separate budget. The Audit Commission (1989) and the National Audit Office (1990) both drew attention to potentially wasteful duplication and overlap of central and local government effort. The development of City Challenge reflected the attempt to move from a programme dominated by large numbers of small schemes in 57 areas towards a programme within which a smaller number of large initiatives would provide the focus for integrated intersectoral and interdepartmental spending programmes. In 1993 the announcement of the Single Regeneration Budget (SRB) represented a further step in expressing this explicit philosophy of addressing the issues of programme fragmentation and 'dis'-integration.

This history of institutional fragmentation followed by the attempt to bring co-ordination and cohesion into the allocation of public expenditure resources raises a number of value for money questions. One important question is whether the combined impact of the various and constantly changing instruments of policy is greater than the sum of the same elements taken separately. One argument might be that the variety of interventions attempted in the 1980s are less representative of the much criticised fragmentation of policy implementation than indicative of a pluralist approach which aims to develop tailor-made and targeted programmes to meet specific needs. Thus Task Forces, UDCs, Estate Action, the Urban Programme, City Action Teams and so on each have their own specific and particular role to play and each needs to be evaluated in the light of the objectives peculiar to the individual programme. Such an argument holds little sway, however, and it is more commonly accepted that the combined and cumulative impact of multiple programmes on cities is more than the sum of the disaggregated parts. Evaluative approaches thus must measure complementarity and synergy and examine the total impact of the separate government programmes. This was the prime focus of the most recent Department of the Environment report on urban policy (Robson et al., 1994) which offered a comprehensive assessment of overall achievement against overall input and argued strongly for the greater integration of policy.

This focus upon the mutually reinforcing role of different initiatives and on the synergistic effects of programme integration highlights the significance of developing coherent and complementary expenditure plans. This in turn directs attention more explicitly onto issues of financial management and control and hence on the evaluation of urban public expenditure within the general context of fiscal efficiency.

Urban Public Expenditure

The DoE's stewardship of public expenditure is dominated by the Local Government block (predominantly Revenue Support Grant and Non-Domestic rate payments) which amounts to a planned £30 billion in 1994–95 and by the Housing block which in turn accounts for £7.5 billion of the planned spend of £39.2 billion. Table 4.1 illustrates the scale of the overall DoE programme and reinforces the important point that reductions in overall grant to local government (and within that to urban programme authorities) dwarf the relatively less significant shifts in urban spending. Nevertheless the Urban and Regeneration Programmes – which include in Table 4.1 the other government departmental (OGD) expenditures transferring to the Single Regeneration Budget – remain the third largest element within the expenditure under overall DoE control and represent much more than a symbolic programme. The words and language of policy may seek to conceal some of the political realities (Edelman, 1977) but the scale and pattern of urban policy expenditure generates a real impact in specific urban areas.

The Urban and Regeneration block at least up to 1993–94 was a volatile as well as rapidly increasing element of expenditure. Annual expenditure changed by significant amounts (up 76 per cent in 1990–91 from 1989–90, up 47 per cent in 1992–93, down 20 per cent in 1994–95). This was in large measure due to the presence of receipts as well as expenditure in the net expenditure figures and to the fluctuating scale of such receipts in the expenditure planning period. In 1994–95, however, regeneration expenditure fell back sharply by over £250 million whilst an even larger fall in spending to £1,263 million is planned for 1995–96.

The proliferation of organisational initiatives throughout the 1980s has already been noted. Important amongst these were the establishment of Urban Development Corporations and the City Grant (former Urban Development Grant) régime, of Task Forces and City Action Teams, together with the phasing out of the 'traditional' Urban Programme. Within recent years the most important single shift has been the introduction of City Challenge, a competitive bidding initiative with 31 winning localities each gaining £35 million over five years and other competing localities receiving nothing. Other changes have included the wind down of the Urban Programme (and the consequent withdrawal of support from a number of areas which over the years have received urban aid), the introduction of a new Urban Partnership scheme (allowing local authorities to utilise capital receipts in conjunction with small amounts of Urban Programme funding), the establishment of an Urban Regeneration Agency (now English Partnerships) to take on responsibilities for City Grant and Derelict Land Grant and for English Estates, and support for the Manchester Olympics bid.

Table 4.1 Department of the Environment spending plans (£million cash)

	88–89	89–90	90–91	91–92	92–93	93–94	94–95	95–96	96–97
Environmental Protection	97	161	220	244	251	267	303	289	285
Countryside and Wildlife	93	98	100	87	118	125	142	148	148
Urban and Regeneration	432	769	1,046	1,092	1,440	1,801	1,543	1,263	1,260
Housing and Construction	2,773	2,923	6,724	7,451	8,178	7,652	7,429	7,583	7,557
Local Government	19,002	19,752	20,527	28,374	31,216	29,406	29,953	30,921	32,502
Departmental Administration	153	182	187	215	223	226	228	230	233
Property Holdings	8	-78	-117	-160	-156	-185	-304	-347	-357
Total DoE	22,558	23,807	28,689	37,303	41,271	39,292	39,294	40,087	41,628
PSA Services			60	50	-69	114	113	64	36
Office of Water Services		2	5	6	7	10	10	8	8
Ordnance Survey	23	19	20	18	20	19	15	13	13
Total	22,581	23,828	28,774	37,378	41,228	39,435	39,432	40,173	41,685

Source: DoE (1994a).

These shifts in institutional structure are reflected in the expenditure programmes within the Urban and Regeneration Programmes segment of the DoE's spending plans (Table 4.2). In the 1993 Plan the composition of the segment differed from previous years in two respects. First New Town Expenditure, hitherto regarded as a separate element of the DoE's programmes, was included within the Urban block (hitherto designated simply as Inner Cities). In practice as a result of a strong programme of land and property disposal the New Towns expenditure line is negative since receipts exceed spending. Lower receipts are thus equivalent to increased expenditure.

Secondly the 1993 Report also included contributions from the European Regional Development Fund within planned public expenditure as a consequence of the application of the principle of increased transparency in relation to the additionality of such expenditure. In the 1994 report Estate Action and Housing Action Trust expenditures were transferred to the Urban and Regeneration budgets (see Table 4.5 below) but to allow comparison with previous years they are omitted from Table 4.2.

There have been, therefore, major shifts in the distribution of the Urban Policy spend between different elements of the programme over the past decade reflecting the primacy of different institutional forms and the political necessity to give new initiatives a dedicated – and in some cases substantial – budget allocation.

Table 4.3 shows the changing share over a decade of the different elements of what was then known as the Inner Cities segment of urban expenditure (i.e. omitting New Towns and ERDF). Within the total the Urban Programme dropped from a three-fifths share to 10 per cent in ten years. UDC spending accounted for an increasing share of spend in the period up to 1992–93. Thereafter a planned decline in UDC spend was offset by the emergence of City Challenge and later by the expansion of the City and Derelict Land Grant programmes now managed by English Partnerships.

The Single Regeneration Budget

In 1993 the Government announced the establishment of the Single Regeneration Budget to be administered through a system of Integrated Regional Offices. SRB brings together twenty targeted programmes and initiatives from four departments with the objective of ensuring co-ordinated planning and management of regeneration activity at the local level. The consequences for public expenditure planning are described in the DoE *Annual Report* (DoE, 1994a) which brings the relevant SRB components of other departments' spending plans into the regeneration segment of the DoE expenditure block for the first time.

Table 4.4 allows a direct comparison to be made between the former presentation of public expenditure on Urban Policy and the new Regeneration Budget. This fulfils an important evaluation requirement in itself in so far as it increases the transparency of public expenditure and makes explicit the changes in expenditure volumes which innovation in presentation often conceals. In Table 4.4, column (i) sets out the planned expenditures in 1994–95 for each of the separate urban and regeneration programme areas. Column (ii) lists those

Table 4.2 Urban expenditure 1988–89 to 1994–95 (£million)

	88–89	89–90	90–91	91–92	92–93	93–94	94–95
Urban Programme	224	223	226	237	236	173	85
City Challenge	–	–	–	–	52	223	214
Task Forces	23	20	21	20	24	18	16
City Action Teams	–	4	8	8	5	3	1
City Grant	28	39	45	41	44	24	–
Derelict Land Grant	68	54	62	77	102	104	–
Urban Regeneration Agency	–	–	–	–	–	34	181
UDCs and DLR	255	477	607	602	515	381	291
Manchester Regeneration	–	–	–	1	13	35	23
Other	–	–	–	–	–	3	4
CF Extra Receipts	–	–2	–4	–7	–6	–4	–
Inner Cities Total	598	815	964	980	984	993	814
New Towns Commission	-558	-463	-333	-372	-169	-95	-123
ERDF	27	8	2	3	27	176	199
Total	67	360	633	611	842	1,074	890

Table 4.3 Share of inner cities expenditure (%)

	85–86	87–88	89–90	91–92	93–94	94–95
Urban Programme	58	51	27	24	17	10
City Challenge	–	–	–	–	22	26
City Grant	5	6	5	4	2	–
Derelict Land Grant	17	16	8	8	11	–
English Partnerships	–	–	–	–	–	22
UDCs and DLR	20	28	58	61	38	36
Task Forces	–	–	2	2	2	2
City Action Teams	–	–	0	1	0	0
Manchester Regeneration	–	–	0	–	4	3
Other	0	0	0	0	1	1
Total	100	100	100	100	100	100
(£ million)	(436)	(483)	(815)	(980)	(993)	(814)

Source: DoE *Annual Reports* 1991 to 1994.

elements which formerly made up the Inner Cities programme, and the total of £814 million is directly comparable to the Inner Cities figure shown in Table 4.2 above. Column (iii) lists those elements which go to make up the Single Regeneration Budget (Hansard, 1994) amounting to the now well known figure of £1,447 million. The addition of the final stage of what was the Manchester Olympics bid related programme, New Towns net expenditure, the ERDF expenditure line, and the Coalfield Areas and Special Grants budgets, together with adjustments for those elements of regeneration activity for which the expenditure is included not in DoE but in other departmental public expenditure plans, produces the overall Urban and Regeneration public expenditure total of £1,543 million (see Table 4.1 above).

In addition to the long-standing 'urban' expenditure programmes and the DoE Estate Action and Housing Action Trust programmes (£373 million and £88 million respectively) the Single Regeneration Budget in 1994–95 incorporates £201 million of expenditure from four government departments other than DoE. Twelve programmes are included, of which grants for services for minority ethnic groups under Section 11 of the 1966 Local Government Act (£60 million) and Business Start Up (£70 million) are the most significant.

The Single Regeneration Budget brings together a range of programmes

Table 4.4   The elements of the Single Regeneration Budget

Public Expenditure	(i) 1994–95 Planned Expenditure	(ii) Former Inner Cities Programmes	(iii) Single Regeneration Budget	(iv) Total
Urban Development Corporations	291			
English Partnerships	181			
		472	472	472
Urban Programme	83			
City Challenge	213			
Task Forces	16			
City Action Teams	1			
		313	313	313
Manchester Regeneration	23			
Other Programmes	4			
		27		27
Estate Action	373			
Housing Action Trusts	88			
			461	461
New Towns (CNT) net	-123			
ERDF	199			
				76
Other Government Departments	201			
			201	201
Coalfield Areas Fund	2			-2
Ministry of Defence	5			-5
Total		814	1,447	1,543

Source : Hansard, 31 March 1994, Col. 918.

but it does not offer increased resources. Indeed as summarised in Table 4.5 the SRB envelope is planned to decrease over the first three years of its existence from £1,447 million to £1,324 million in 1996–97 (Hansard, 1994). This represents a cut of £123 million (over 8 per cent) from 1994–95 levels and indeed between 1992–93 (the year in which SRB equivalent programmes peaked at £1,614 million in cash terms) and 1996–97 the SRB will have been reduced by almost £300 million. In constant 1994–95 prices this amounts to a cut of 18 per cent (Hansard, 1994).

Table 4.5 The Single Regeneration Budget 1993–94 to 1996–97 (£ million cash terms)

	1993–94	1994–95	1995–96	1996–97
Expenditure on Programmes within the SRB	1,614	1,447	1,332	1,324

Although presented as a single budget the SRB is in practice preallocated to specific programmes, and top slicing for Urban Development Corporations, English Partnerships and Housing Action Trusts accounts for £560 million of the 1994–95 budget. Additional commitments – to City Challenge and to continuing Estate Action schemes – make further inroads into the budget and only some £100 million is likely to be available for 'new' bids in 1995–96. The issue of pre-commitment becomes more significant as the programme reduces because although in principle the falling away of commitments releases resources, in practice cutbacks in the overall budget mean that fewer new projects can be funded.

The reductions in funding are, however, to be experienced unevenly. UDC spending is expected to fall by some £40 million between 1994–95 and 1996–97 but Housing Action Trust expenditure will remain level and English Partnership expenditure will increase by £30 million. The reduction in resources will thus be largely carried by the remainder of the programme – the programmes of other government departments and the programmes involving schemes initiated by local government. Of the reduction of £133 million identified above, £120 million will be carried by these latter programmes.

The Single Regeneration Budget is thus far less flexible than initially suggested. It represents less a renewed localism than a realigned centralism (Stewart, 1994a) reflecting a resource shift towards centrally controlled institutional forms of regeneration and tighter expenditure management systems. The reducing level of total resource together with the top slicing to give priority to specific programmes has important implications for the future management of public expenditure. For evaluation of the efficiency and effectiveness of the new regime in financial terms a number of consequences emerge.

First is the significance of the severe competition which will exist for resources from SRB. Local authorities and/or TECs (or whatever local coalitions of interest put together bids) will be fighting for relatively modest sums of money in the early years of SRB and only a fraction of the bids submitted will be likely to succeed. The effect will be to pitch locality against locality, within and between regions. The incorporation of private and non-statutory sectors into the building of competitive local growth coalitions will further dilute the will and the capacity of local authorities to use SRBs as the vehicle for expressing local political (and social) priorities. Conversely English Partnerships, exempt from competition for resources, with enhanced budgets, and with a remodelled development grant regime will exert strong influence through its six (*sic*) regions on the shape of regeneration activity over the coming years. Similarly Housing Action Trusts represent chosen instruments for the delivery of housing policy creating autonomous tenant-based institutions independent of local authority control.

Within this tight competition and in the face of resource scarcity successful bids are thus likely to be those with multi-sector involvement (private and/or non-statutory) together with the support of English Partnerships. Leverage will remain important with many bids dependent upon joint public/private sector financing. This reliance upon private sector leverage echoes strongly the tone of the urban initiatives of the 1980s and early 1990s and emphasises the interdependency of public and private policy. It also re-emphasises the vulnerability of urban policy to the development market. If even the limited resources available within the new SRB are to be effectively used the programme as a whole must be managed in a manner which makes it immune to the impact of cyclical movements in the development industry. Since the expenditure within SRB depends in part on the pace of resurgence in the market and hence upon the flow of receipts to Development Corporations the relationship between public initiatives and private funding is crucial.

It is at this point that evaluation of the experience of recent years is relevant. How far does the planning and management of public expenditure during the development boom and downturn of the late 1980s and early 1990s provide confidence in the ability of the institutions of urban policy to make the best of scarce SRB resources? The following section addresses this question using material from the DoE's annual public expenditure reports. It focuses on the City Grant and Derelict Land Grant regimes and on the Urban Development Corporations and on the New Towns programme. All are in different ways closely linked to the development sector and their financial performance is a function of the extent to which public plans can be maintained in the face of cyclical movements in the development market.

Interdependence in Public and Private Resource Provision[1]

City Grant

In November 1993 responsibility for City Grant was transferred to English Partnerships. City Grant has been paid to private sector developers to support capital projects for which the gap between development cost and the value on completion of the project is such that the development would not otherwise be undertaken. Over its life £300 million of City Grant funding relating to 360 projects levered in nearly £1.3 billion of private sector resources. Peak expenditure (£45 million in cash terms) occurred in 1990–91.

Two features of the City Grant regime in the early 1990s stand out in the context of the downturn in the development market. First was the inability to spend planned levels of grant. For the two-year period 1991/92 and 1992/93 as a whole the expected outturn expenditure (£85.1 million) as expressed in the DoE 1994 Report is £56 million below that planned for the same two-year period in 1991 (DoE, 1991). Even for the seven-month period within 1993/94 the expected outturn amounts to only a third of that planned only a year earlier.

Secondly (Table 4.6) the ratio of private investment to grant fell in 1991/92 and 1992/93 as compared with earlier periods (i.e. the leverage ratio fell). The mix of schemes (housing, environmental, industrial) changed over time, as did the regional distribution of projects, and these affected the leverage ratio. Nevertheless a fall in the leverage ratio from 4.48 (for the period 1988–91) to 3.94, and a further fall between November 1992 and November 1993 to 3.18 reflected the difficulties of levering in private investment and implied a level of grant expenditure £19 million greater than that which would have been required to generate an equivalent private sector investment between 1988 and 1991.

Table 4.6 Public sector leverage in City Grant projects

Period	Private Sector (£ million)	Grant (£ million)	Leverage Ratio
May 1988 – October 1991	919	205	4.48
May 1988 – November 1992	1,163	267	4.35
May 1988 – November 1993	1,284	305	4.21
October 1991 – November 1992	244	62	3.94
November 1992 – November 1993	121	38	3.18

Derelict Land Grant has also been transferred to the management of English Partnerships. It has been available to a range of public, private, charitable and non-statutory landowners for the return of derelict land to beneficial use. Expenditure could be directed to 'hard end' use (development of land for housing, industry, commerce) or 'soft end' use (environmental, agriculture, forestry, recreational). Almost 60 per cent of the land reclaimed in 1990/92 was for hard end use. DLG was directed mainly to local authorities; 93 per cent of the 1992/93 expenditure of over £100 million was expected to be on local authority projects. Grant was paid gross to local authorities (with increased value recovered when land is disposed of) but was paid net of the increase in land value to non-local authority land schemes.

In contrast to City Grant the impact of the decline in the development market on Derelict Land Grant was much less visible with outturn figures remaining relatively consistent with the planning figures of earlier years. This is because DLG has been directed predominantly towards local authority schemes (with private sector involvement being in consequence much less significant), and because disposals (and receipts) made a smaller contribution to net public expenditure. The conclusion is that the DLG policy instrument has been a much more stable and predictable route towards the bringing forward of land for development than has been City Grant precisely because the parallel involvement of private sector resources is a less significant policy objective.

Urban Development Corporations

Urban Development Corporations (UDCs), widely perceived as the flagship of urban policy in the 1980s, were set up in four tranches from 1981 onwards. The National Audit Office (1993) has offered confirmation of the general vulnerability of UDCs to the property market cycle and of the consequence of their dependence on partnership with the private sector. NAO offered a number of observations:

> Economic recession, problems of land assembly, and changed decisions by developers on the timing and extent of progress have all delayed progress on regeneration and called for some re-working of strategies.

> The economic downturn has occurred at a crucial stage in the strategies of the UDCs . . . has significantly reduced confidence in private sector development, delayed progress or cut back the scale of a number of projects, and restricted the scale of receipts generated by UDCs from site disposals.

More explicitly, the NAO confirmed the fall in private sector leverage ratios, commenting that for UDCs ratios fell from an average of 5.2:1 in March 1988 to 3:3.1 in March 1992. Ratios for individual urban areas vary from 1.8:1 to 6.1:1 reflecting not only the stage of completion of UDC plans but also the differential impact of the recession.

In terms of the predictability of expenditure and receipts, the experience of UDCs is of extreme volatility with regular and significant annual adjustments having to be made to both planned and actual expenditure figures as the development cycle unfolded. Table 4.7 shows the successive planning and outturn figures from DoE Reports (planning figures being shown in italics). The series of annual reports shows how regular revisions to both expenditure and receipts are required as actual outturn is affected by the development market.

For the four-year period 1989/90 to 1992/93 (years for which accurate outturn figures are now known) the difference between the original planning figures (DoE, 1991) and the actual gross expenditure incurred (DoE, 1994a) amounted in cash terms to £446 million (22 per cent of the original planned figures). As far as 1993–94 is concerned it had been anticipated that gross expenditure by UDCs would be reduced by almost a quarter of a billion pounds (from £621 million to £387 million) between 1992/93 and 1993/94.

The latest figures suggest, however, that the reduction will be some £100 million less than planned. Estimates of likely receipts have also been extremely volatile. In the 1993 Report the level of planned receipts for 1993/94 was reduced by 50 per cent from that anticipated in 1992; in the 1994 Report planned receipts for 1993/94 were doubled – back to the 1992 prediction. Taking gross expenditure and receipts together the effect on net public expenditure in 1993/94 is thus an increase of only (sic) £44m. between the plans of the 1993 Report and the estimated outturn of the 1994 Report. An upwards adjustment of £101 million to the expenditure figure is offset by an increase in receipts of £57 million.

Table 4.7 Urban Development Corporations expenditure and receipts

Gross expenditure (£ million)

	88–89	89–90	90–91	91–92	92–93	93–94	94–95	95–96
1991 Annual Report	339	478	616	541	410	375		
1992 Annual Report	338	513	608	574	481	423	400	
1993 Annual Report	359	554	663	674	621	387	368	349
1994 Annual Report	359	554	662	657	618	488	na	na

Receipts (£ million)

	88–89	89–90	90–91	91–92	92–93	93–94	94–95	95–96
1991 Annual Report	104	42	66	68	105	72		
1992 Annual Report	104	77	55	72	149	100	95	
1993 Annual Report	104	77	55	72	101	50	75	65
1994 Annual Report	104	77	55	55	103	107	na	na

Net public expenditure (£ million)

	88–89	89–90	90–91	91–92	92–93	93–94	94–95	95–96
1991 Annual Report	234	436	550	473	305	303		
1992 Annual Report	234	436	553	502	332	323	305	
1993 Annual Report	255	477	608	520	520	337	293	284
1994 Annual Report	255	477	608	602	515	381	291	254

Three main observations emerge from this discussion of UDCs. The first relates to the frequency and scale of the adjustments made to public expenditure planning figures over the years, together with the continuing and regular difference between planned expenditures and receipts and the outturns which are reported in subsequent years. Secondly comes the volatility of planned and actual receipts, and the impact of receipts on the net expenditure of UDCs (recognising that the cycle may involve only a postponement rather than an absolute loss of receipts). A third observation concerns the relentless increase up to 1993/94 in the planned gross and net public expenditure requirements of the UDCs.

New Towns

Although New Towns expenditure is not within the SRB the New Town experience illustrates further the impact of development sector recession on public expenditure management. All the former New Town Development Corporations have been wound up and their assets transferred to the New Towns Commission. Expenditure (other than a small housing repair element) is now concentrated upon investment in development sites in preparation for their disposal, and in practice the most significant contemporary element of the New Towns programme is the management and disposal by the New Towns Commission of the residual assets of the development corporations. Since 1979/80 disposals have realised over £4,000 million, and new investment has been concentrated on activities designed to improve the viability of remaining development sites.

Receipts for the 1991/92 to 1993/94 period (DoE, 1994) are expected to be £631 million less than envisaged three years earlier (DoE, 1991), and in the two-year period 1992/94 to be £351 million less than predicted even in the 1992 Report. In each of the last two years, therefore, the financial impact of the property market downturn on New Town net expenditure – and hence on the DoE urban expenditure block as a whole – is the equivalent of two dozen City Challenge initiatives. Future receipts are expected to continue to be lower than in earlier years and reduced expenditure is acknowledged only to 'offset in part the reduced level of receipts in the current economic climate'. The New Town figures illustrate graphically the impact of the downturn in the property market but while the Annual Report explicitly acknowledges this impact there is little evidence of exploration by DoE of the policy or practice implications of the development cycle for disposal policy or for property holding.

Conclusions

For a number of years the growing dependence of urban regeneration on policies and practices of physical redevelopment has been recognised (Stewart, 1987; Healey, 1991). The bulk of the institutional framework and many of the policy instruments (UDCs, grant regimes, etc.) have been predicated upon the assumption that private sector resources can be levered into the inner city and that the development process will engender lasting regeneration. It is clear from UDC evaluation, from the City Grant and DLG data, and from the New Towns

Commission that policies which are designed at least partly to respond to market failure or at least to fill gaps in market mechanisms are themselves vulnerable to market forces in respect of their implementation. This fragility has been most evident in recent years and is well illustrated by the trends in public spending detailed above.

Urban expenditure has been vulnerable to the impact of cyclical movements in the development market and that expenditure has been both volatile and unpredictable. In 1994 outturn City Grant expenditure over the three years 1991/92 to 1993/94 was identified as being some £85 million greater than the planned levels foreseen in 1991; over the same three years UDC expenditure was over £400 million above earlier planning levels; net New Towns expenditure over the same three-year period (largely negative receipts) was over £500 million more than anticipated in the public expenditure exercise three years earlier. These three programmes incurred net public expenditure £1 billion above what had been anticipated three years earlier. The fragility of the public expenditure planning process in the face of turbulence in the development market is clear. This analysis highlights the predicament of public expenditure management together with a confusion of philosophy as to how to respond to cyclical shifts within a policy which is essentially market driven. The dilemmas are not new (Barrett, Stewart and Underwood, 1978) but have been reinforced by the severe impact of the latest cyclical downturn.

In practice we see programmes where political imperatives override functional efficiency to the detriment of both proper public administration and political achievement. What was initially a policy driven by political ideology (the merit of private sector leadership, reward to capital and an attack on public sector bureaucracy), allied to predominantly financial objectives (reduction of public spending by leverage of private sector funding), was insufficiently specified to meet the needs of urban regeneration at a time of cyclical downturn. Even DoE now recognises that:

> experience has shown that regeneration strategies relying on a few flagship projects are vulnerable to planning difficulties and uncertainties in the property market.
> (DoE, 1994b)

There are two observations of relevance to the future management of urban policy to be drawn from this evaluation of past expenditure trends. One relates to the emergence of the Single Regeneration Budget and its standing within Whitehall.

Thain and Wright (1990) identify both 'prudential' and 'contingent' expenditure as being part of the Treasury's coping strategy for the management of expenditure in the face of relentless pressure for increased public resources. In the urban policy context prudential arguments would include the need to protect the cities on grounds of social control let alone electoral benefit from the worst effects of expenditure cuts, and to offer a symbolic programme to engage local governments whilst at the same time conceal other cuts in subventions to localities. City Challenge represents such expenditure par excellence. Contingent expenditure – the prime focus of this chapter – has been necessary because recession in the property market has left the public sector with a larger job of private sector leverage than had been anticipated.

The successive and spectacular adjustments to public spending plans consequent upon shifts in the profile of both expenditure and receipts will now be contained within the Single Regeneration Budget. It is reasonable to suppose that at least one of the reasons which induced the Treasury to agree to the SRB was the assurance that any overspend in a programme within it would be contained within the overall total. Thus 'contingent' expenditures would become a thing of the past as virement within the overall SRB responds to unanticipated shifts in particular elements within it. This internalisation of under or overspend within the SRB accentuates the vulnerability of non-development elements of the single budget to cyclical fluctuation. If for example the UDCs fail in 1995–96 to realise their planned level of receipts the shortfall in public expenditure resources may need to be found elsewhere in the SRB.

Supply side support has been a feature of spatial policy for decades (advance factory building being the cornerstone of much regional intervention) but it has been rare until recently for public resources to be used with such determination to sustain the development function through the private sector. Healey (1991) identifies the 'temporal variability' of the conditions which in different urban areas could justify such supply side support but the analysis of aggregate urban policy expenditure suggests limitations in the ability of DoE to reflect the temporal as well as the geographical variability of the development market.

The implications from this analysis for the development of a more effective policy thus reflect the need for a much greater degree of disaggregation and refinement in the management of government programmes in terms of their relationship to local or regional development property markets. Variation in the performance of such markets clearly exists and is well documented (Nabarro and Key, 1992; Investment Property Databank, 1993). Retail, office and industrial markets differ not simply in terms of the relative mismatch of supply and demand in different parts of the country but also in the pace at which they are likely to emerge from the property downturn. Performance across the country varies. Analysis of such market information needs to be allied to a mid-1990s understanding of the behaviour of the varied players now engaged in property and development markets. The next few years will see a continuing interaction of structure and agency (Healey and Barrett, 1990) which will provide a complex context for the management of urban public expenditure.

English Partnerships has a crucial role to play in any improvement in the management of public expenditure on urban policy. Their potential contribution to improved management of the interdependence between public and private sectors stems not solely from their own significant regeneration budget (£220 million in 1996–97), but also from the influence and leverage that they will bring to bear on other programmes – not least through their linkage with the Government's inward investment programmes. With a regional perspective and a new grant regime to replace City Grant and Derelict Land Grant, English Partnerships ought to have the capacity to introduce a greater degree of sophistication into the implementation of development related urban policy.

It will be important that adequate advice on the implications of the development market is given to the new Ministerial Committee on Regeneration since whilst that Committee will have an overview on regeneration programmes of business support and infrastructure, it will nevertheless be top slicing the Single Regeneration Budget in 1995/96 to the extent of at least £0.75 billion in order to provide funding for English Partnerships, Urban Development Corporations, City Challenge and Housing Action Trusts (DoE, 1994a). SRBs purport to offer a realignment of urban policy in England by providing greater flexibility and local responsiveness. Volatility on the scale experienced in recent years would seriously undermine the potential of the new SRB to offer flexibility and effectiveness in resource management. If the Government's expenditure planning and management capacities, whether in Whitehall and/or in the integrated regional offices, are not improved the impact on the implementation of the new Single Regeneration Budget may well be dire.

Note

1. This section of the chapter draws upon material more fully developed by the same author in a recent article in *Public Money and Management* (Stewart, 1994b).

5

URBAN POLICY AND URBAN EVALUATION: THE IMPACT OF THE EUROPEAN UNION

Michael Chapman

Introduction

The purpose of this chapter is to discuss the growing influence of the European Union (EU) on urban policy in the United Kingdom. In a period of increasing economic, environmental and social stress in our cities many UK local authorities and local economic development agencies in the larger metropolitan areas have sought to utilise the financial resources available from the European Commission (EC). The attention given by policy-makers to European policy and programmes occurs at a time when UK central government expenditure and resources on urban policy are being reduced and redirected, while the European policy agenda is increasingly addressing the specific problems experienced by urban environments. The lack of vision and cohesion in urban policy has been consistently identified as a serious constraint on the development of suitable policies for urban areas (Lawless, 1991). While on the one hand there has been a growth in the multiplicity of urban policy instruments and strategies there has also been a simultaneous drive towards greater efficiency and cost effectiveness in the delivery of urban policy over the past fifteen years. The result has been the need for close examination and evaluation of individual policy instruments and different urban regeneration strategies (Robson *et al.*, 1994; Imrie and Thomas, 1993; Campbell, 1990). There remains, however, a need for evaluations to take into account the impacts of local, national *and* European policies on the urban environment. This chapter contributes to this debate and tries to establish the significant role made by the different policy instruments of the EU and how they impact on urban areas. It is necessary to recognise that Europe is becoming a key component in the formulation of urban regeneration strategies for many British cities. As a consequence those involved with the evaluation of urban policy should now recognise the growing influence of the European policy agenda and seek to ensure that the contributions made by European funding opportunities are clearly accounted for.

In keeping with the principle of subsidiarity, as agreed by the ratification of the Treaty of Maastricht in February 1992, policies designed to address and tackle urban problems are most appropriately undertaken by the individual member state or the relevant regional or local authority. Although 80 per cent of Europe's population (i.e. some 250 million people) live in urban areas the European Commission has no direct competence for urban affairs. Those who live in the city often enjoy the advantages of city life but they also have to put up with the disadvantages of the so-called 'urban condition', or in other words the heavy economic, environmental and social burdens which our cities suffer (European Foundation for the Improvement of Living and Working Conditions, 1994, p. 24). As yet there is no integrated framework for urban policy at the European level as there is for transport, the environment or for the reduction in regional disparities. Yet although the Treaties of the EU do not provide a mandate for the Commission to develop a European-wide urban policy it is becoming increasingly evident that urban areas are benefiting both directly and indirectly from the policies and programmes of the EU. The growing recognition of the important economic and social role played by cities and urban regions in Europe is reflected in reports published by the Commission (CEC, 1990; 1991; 1992a). But of even greater importance has been recent EU legislation and research which has continued to identify new roles and spheres of activity for the EU in combating economic, social and environmental problems which confront European cities (CEC, 1990; 1993b; 1993c; The European Foundation for the Improvement of Living and Working Conditions, 1994).

One of the most significant moves towards the development of an explicit European urban policy has been the recent introduction of a new Community Initiative (CI) in the field of urban policy. The initiative, known as URBAN, aims to tackle economic, social and environmental problems in urban areas (CEC, 1994b) and supplements the existing efforts by the Community to address urban deprivation and disadvantage. In addition to financial support already provided for cities under Objective 1 and Objective 2 regions, this new initiative adds an extra dimension to Community support for urban areas. The introduction of this initiative cannot hope to counter the range and severity of problems facing European cities. It instead should act as catalyst in a broad based approach and complement existing national and local urban policies.

In the post-Maastricht Europe attention is still focused on the need to address significant regional disparities throughout the Community. However, there is a growing awareness of the ramifications of urban problems and the profound differences between European urban areas. Above all urban problems require specific solutions and specific plans which are best initiated by local and or national governments. The European Commission does not see itself as a distant organisation making plans and presenting solutions in a centralised manner. But due to the scale of these problems there is ample justification for European involvement jointly with the relevant local and national organisations. As the European influence continues there is a need for further analysis not only of the impacts of European policy but also of the direction of future European policy in urban affairs.

This chapter will begin by highlighting recent EU and EC concern for urban areas including an increasing linking of social cohesion and exclusion to the condition of cities. It will then review the implications of current EU programmes for urban areas, before suggesting that we can now speak in terms of a 'Europeanisation' of urban policy in the UK.

Diversity and Change: European Cities in Transition

Since the early 1980s the EC has been interested in urban change, and monitoring the significant structural, demographic, economic, social and environmental changes that have taken place in many European cities. In 1990 the Commission produced its Green Paper on the Urban Environment (CEC, 1990). The Green Paper identified the range of problems facing European cities and suggested that the (then) Community could take certain lines of action in areas such as transport, historical heritage, environmental heritage, industry, energy management, urban waste and urban planning. While the Green Paper was mainly concerned with urban environmental quality it nevertheless represents an important step towards the establishment of a European-wide urban policy. For the first time the Commission broached the issue of whether the Community wanted to extend the levels of financial support for urban regeneration and environmental improvement over and above levels as allowed through the operation of the Structural Funds (see later). Understanding the impact of European-wide spatial change was enhanced by the publication of the report *Europe 2000 – Outlook for the Development of the Community's Territory* in 1991 (CEC, 1991). This report argued that Community policies were having, and would continue to have, a significant impact on the land use and physical planning of urban areas. This view was reinforced by the recent publication of a report by the Royal Town Planning Institute on *The Impact of the European Community on Land Use Planning in the United Kingdom* (The Royal Town Planning Institute, 1993).

The objective of the *Europe 2000* report was not for the Commission to produce a master plan for the spatial development of the Community but instead the publication was seen as a necessary term of reference for future European spatial planning. The report concluded by making key observations concerning the functioning of the European urban system:

> The majority of the larger urban regions have been passing through a cycle of urbanisation or centralisation of employment and population into the central city. This has been followed by a process of suburbanisation to the hinterland and then – for many of the cities of northern countries in the Community – a phase of relative or absolute decline in the central city or even the entire urban region. Many of the cities of southern Europe are still in the early stages of the cycle. Some of the northern cities have passed through the entire cycle and there is some evidence of a re-centralisation of population and employment in certain sectors taking place.
>
> The later stages in the process are associated with a more general development of a more balanced or decentralised urban system as growth has been strongest in smaller and medium-sized cities, compensating for the decline of the larger cities.
>
> In parallel with these changes there have been the internal changes within cities, notably the persistence of areas of severe poverty and poor social and physical environment, whether in the inner city areas or the peripheral areas of social housing or unplanned immigrant housing. In many instances, these areas are the homes of ethnic minorities or other recent immigrants from overseas or Central and Eastern Europe.
>
> (The Royal Town Planning Institute, 1993, pp. 89–90)

Changes to the European urban system are associated with established economic, social and demographic factors. As these factors persist the urban system will continue to respond and adapt. As there is no single functional hierarchy of cities or urban regions in Europe the relative position of a city will thereby shift according to the relationship it has with other cities with which it is in competition for economic resources (RTPI, 1993).

Several other background reports to *Europe 2000* were published by the Commission. The Directorate General responsible for regional policy in the Commission, DG XVIV, commissioned a study on *Urbanisation and the Functions of Cities in the European Community* (CEC, 1992a), which analysed the competitive position of European cities by means of 24 European city case studies. In the UK the cities used for the study included Glasgow, Liverpool, Birmingham and London. It argued that a factor to explain the changing function of European cities was the importance given to the creation of the single market. As the SEM comes into operation successful cities would be those urban environments which had the ability to compete in the conditions established by the single market. This would be achieved by the diversification of local economic structures, the ability to attract inward investment and the servicing of new markets throughout the whole of the Community. A small handful of dynamic and resourceful cities will continue to flourish: examples included Hamburg, Rotterdam, Dortmund, Montpelier and Seville. Those cities which failed to adapt to the competitive pressure introduced by the SEM would increasingly become uncompetitive in the European urban hierarchy. These cities would be characterised by severe structural weaknesses in their urban economies with inadequate physical infrastructure and communications networks for the new economic activity created by the single market. Often located on the periphery of Europe these cities have been unable to break away from their industrial legacy and include cities like Marseilles, Dublin and Naples (CEC, 1992a).

Notwithstanding the position of individual cities the Commission has identified a growing trend towards the spatial concentration in urban areas of high unemployment, the political and physical segregation of economically marginal groups, poor condition and provision of social housing, fiscal stress, transportation and environmental problems and growing ethnic populations. While many cities are engaging in competition and entrepreneurial activities as a response to these problems the approach adopted by the EC emphasises the need for greater co-operation between cities and not competition. Above all there has been a growing concern at the significant levels of poverty and deprivation being concentrated in urban areas. As Mega (1993) comments: 'European cities are not only the centres of economic, social and cultural activity but they represent and reflect the economic and social disparities within European society and the marginalisation of certain disadvantaged groups in that society' (Mega, 1993, p. 527).

Widening the Debate: the Environmental and Social Dimension to Urban Integration and Cohesion

Previous debates over the need for a European urban policy have tended to focus on the European urban system and how individual cities have coped with the economic restructuring, globalisation of production, growing environmental awareness and the economic and social uncertainties which are a consequence of the SEM. But significant policy changes at a European level are occurring which will have profound implications for the future of urban policy in countries like the United Kingdom.

The development of an integrated European environmental policy has important implications for the enhancement of new sustainable urban management policies. The Fifth Environment Action Programme provides the basis for current thinking on European environmental concerns. It represents a further step towards the prudent and rational utilisation of resources and reflects the promotion of sustainable growth as a major policy objective as outlined in the Maastricht Treaty. The European environmental policy agenda therefore has significant implications for local authorities and community groups developing strategies in urban areas.

During a period of continuing European integration a question arises over the economic and social integration of disadvantaged groups in the population and in particular the integration of such groups in urban environments. The EU cannot be developed without internal cohesion, and the future of European integration depends on the development of a social and environmental dimension to Community-wide policies. The risk for individual member states is that the economic, demographic and social pressures will continue and further add to the polarisation of the urban population. The Commissioner for Employment and Social Affairs, Mr Padraig Flynn, reflected the concern of the EC when he stated that:

> Europe faces many challenges, but one of the most difficult, which emerged in the 1970s and 1980s, is the changing nature and increasing incidence of poverty and deprivation. Whereas the Community entered the 1990s with a great deal of confidence, mainly on the foot of the prospects offered by the Single Market, at the same time, there were growing signs that the social situation of Europe's citizens might not be making equally positive progress. The effects of the world recession have intensified fears that social exclusion – a complex blend of interrelated factors – could become a persistent feature of the European Social Landscape.
>
> (CEC, 1992b)

Industrial decline and economic restructuring in the inner city areas and the outer estates of many European cities have resulted in the combination of physical and social decline with high concentrations of unemployment in specific urban areas. The term which is often used by the Commission to describe this problem is 'social exclusion'. The notion of social exclusion refers to the multiple and changing factors which result in people being excluded from the normal exchange, practices and rights of modern society. Poverty is one of the most obvious factors which can indicate social exclusion, but social exclusion also refers to inadequate rights in housing, education, health and access to services. Social exclusion affects individuals as well as certain groups in society. It can be identified in urban areas and rural areas alike, as people are in some way subject to discrimination or segregation. These conditions when combined with low income levels have produced a series of problems which can be summarised as follows:

(i)   Lack of access to amenities, facilities and services which should be available,

(ii) Lack of mobility to jobs and career opportunities which would improve social position,
(iii) Lack of residential mobility to areas with better prospects,
(iv) Lack of choice in buying goods and services,
(v) Not receiving due health care, social support and education.

(The European Foundation for the Improvement of Living and Working Conditions, 1990, p. 95)

Data to illustrate these trends is problematic at the level of the EU. Each member state is affected by social exclusion but the attributes and the form of exclusion mirror the precise problems experienced by different countries and particular urban areas. The most recent comparable figures for the whole of the Community relate to 1985, and most observers would comment that since then the position across the Community and in individual member states has become much worse. In 1985, in the twelve member states of the Community:

• 50 million people were poor,
• 14 million people were unemployed, of which more than half were out of work for at least one year, and about a third were out of work for at least two years,
• 33 per cent of the long-term jobless had never worked,
• 35 per cent of the unemployed were less than 25 years old,
• 18 per cent of those aged under 25 were unemployed,
• 3 million people in the Community were homeless.

(*Source*: Eurostat Rapid Reports, Population and Social Conditions, 1990. The definition of poverty is a person living in a household for which the disposable income per adult equivalent is less than half the average disposable income per adult equivalent in that member state. In CEC, 1993e, p. 1.)

With the signing of the Treaty of Union Europe enters a new phase of policy development. The Treaty placed a strong emphasis on the importance of achieving economic and social cohesion as a prerequisite for the development of the Community as a whole. One of the most important challenges for the EU is to bring about the economic and social cohesion between the poorer regions and the richer parts of the Community. This effort is required if the poorest countries in the Community, Spain, Portugal, Greece and Ireland, are going to be able to take part in the final phase of economic and monetary union. This commitment towards achieving economic and social cohesion is reflected in the Community's Agricultural, Regional and Social Funds and through the establishment of a Cohesion Fund which was expressly set up for this purpose.

More recent policy documents – including firstly the White Paper *Growth, Competitiveness and Employment;* secondly, *The Presidency Conclusions*, the document which presents the actions to be taken following the White Paper and the Green and White Papers on European Social Policy – indicate the need for further action to be undertaken in order to create new employment opportunities across the member states in order to reduce the overall level of unemployment in the Community. Persistent high levels of unemployment are also considered to act as a barrier towards meeting the requirements for future economic and monetary integration. The promise of further action to create new employment opportunities and reduce the levels of European unemployment are likely to have important implications for urban areas. The views expressed in these documents signal a new policy agenda emerging from Europe and in many ways reflect the Commission's position on social exclusion as outlined in the communication *Towards a Europe of Solidarity: Intensifying the Fight against Social Exclusion, Fostering Integration* (1992b). This document expresses concern at a European level for the need to identify solutions which tackle the problems of social exclusion which are often associated with deprived urban environments. The difficulty that the Commission is confronted with is that any future European policy on social exclusion has to take into account the level of competence, the resources available and actions which are in accordance with the principle of subsidiarity.

Europe's Influence on Urban Areas?

If action in relation to social cohesion lies mainly in the future, this section outlines some of the major impacts EU programmes have already had on urban areas.

Many UK cities, such as Manchester, Liverpool, Cardiff, Belfast and Glasgow, to name a few, are directly involved with different Community programmes, which make a positive contribution towards the improvement in living and working conditions in urban areas. Local authorities can pursue a number of potential sources of EC funding and assistance, but in overall terms the most significant policy instrument has been the operation of the Structural Funds. The Structural Funds are the main Community funding mechanism to combat regional disparities. In the period 1989–93, the Structural Funds have contributed 63 billion ecu (at 1989 prices) to policies and programmes intended to assist programmes for investment, restructuring, training and employment. The Structural Funds are not the only source of funding available. Urban areas have also benefited from certain Community Initiatives aimed at employment training, the Community's anti-poverty programme, the Financial Instrument for the Environment (LIFE Programme, which ties in with the EC Fifth Action Programme relating to the Environment, *Towards Sustainability*) and opportunities for cities and urban areas to participate in European networks.

As the influence of Europe becomes more significant many local authorities have been quick to realise the potential of funding from the EU. Cities like Sheffield, Birmingham and Glasgow have dedicated officials and departments with responsibility for monitoring European affairs and keeping up to date with changes in European legislation and funding. All this activity does not mean that gaining funding from the EU is an easy option. Obtaining funds requires well thought out planning, good quality bids which reflect an understanding of current European priorities and regulations, and above all the requirement for matching funding. But as expenditure on urban policy in the UK has been reduced it is not surprising that urban local authorities have looked towards Europe in order to fund projects.

With the completion of the SEM and the ratification of the Maastricht Treaty, a proposed revision of the Structural Funds has been undertaken for the period from January 1994 up to the end of 1999. Although the basic objective of the Structural Funds is to assist economic and social cohesion through the reduction of regional disparities, the reform (CEC, 1993a) does in fact pay attention to the specific problems experienced by urban areas. The EC recognised how urban areas had benefited from Community action during the previous financial term. But for the period 1994–99, a more ambitious and co-ordinated approach is being introduced by the Commission. This approach is being implemented through the continuation and strengthening of the financial support via the Structural Funds to cities within Objective 1 and Objective 2 regions, through the support for the Urban Pilot Projects and city networks under the RECITE initiative; by the launch of a new Community Initiative for urban areas (URBAN), which is aiming to promote innovative actions which can be used as examples to be diffused in cities across the EU, and promotion of networks of exchange of experience and co-operation; through the development of further research into urban problems supported by the new Fourth Research Framework programme; ensuring that an urban dimension is introduced to other Community policies, especially those relating to the urban environment; and strengthening the dialogue between cities and their representative organisations in the newly established Committee of the Regions. The Commission is likely to monitor the impacts of Community action on urban areas during this next phase of European policy development. The argument is just as significant for those policy-makers and researchers who are evaluating the outcomes of different policy initiatives in urban areas. To be comprehensive in policy evaluation there is a need for a better understanding of the different European policy instruments which impact on our cities.

European Policy Instruments
The Structural Funds

The reduction of regional disparities is considered to be one of the principal tasks of the EC and the mechanism used to do this task is collectively known as the Structural Funds. The key components of the Structural Funds include:

- European Regional Development Fund (ERDF): ERDF supports environmental projects, investment in infrastructure and assistance to small and medium sized enterprises.
- European Social Fund (ESF): the ESF supports training, retraining and vocational guidance with particular emphasis on young people and the long-term unemployed.
- European Agricultural Guidance and Guarantee Fund (EAGGF): EAGGF supports measures to speed up the adjustment of agricultural structures with a view to the reform of the Common Agricultural Policy.
- Financial Instrument for Fisheries Guidance (FIFG): FIFG supports measures to speed up the adjustment of fisheries structures with a view to the reform of the Common Fisheries Policy.

Table 5.1 indicates the main changes that have occurred through the revision of the Structural Funds for the period 1994–99.

The resources allocated for the Structural Funds for the period 1994–99 will amount to 141,471 million ecu (CEC, 1993a). Of this 70 per cent will be accounted for by the Objective 1 regions. The spatial coverage

Table 5.1   Priority objectives for Structural Funds

Former regulations 1989–93	Revised regulations 1994–99
The 1988 reform had established five priority objectives for the Structural Funds.	The revised Regulations do not amend the definition of Objectives 1 and 2.
Objective 1: development and structural adjustment of the regions whose development is lagging behind.	The new Objective 3 combines the tasks of the current Objectives 3 and 4 and also aims at 'facilitating the integration ... of those threatened with exclusion from the labour market'.
Objective 2: converting the regions or parts of regions seriously affected by industrial decline.	
Objective 3: combating long-term employment (more than twelve months).	The new Objective 4 must give effect to the new tasks laid down for the ESF in the Maastricht Treaty: 'to facilitate workers' adaptation to industrial changes and to changes in production systems'.
With a view to the reform of the common agricultural policy:	Objection 5 aims to promote rural development:
Objective 5a: speeding up the adjustment of agricultural structures;	Objective 5a maintains its initial goal of speeding up the adjustment of agricultural structures as part of the CAP reform, but it also includes aid to modernise and restructure fisheries;
Objective 5b: development of rural areas.	
Objectives 1, 2 and 5b are specifically regional in nature; they involve measures restricted to certain eligible regions or parts of regions.	Objective 5b facilitates the 'development and structural adjustment of rural areas'.
Objectives 3, 4 and 5a on the other hand cover the whole of the Community.	Moreover, in accordance with the conclusions of the Edinburgh Council of 1992, the revised Regulations make provision to include areas which are suffering from a decline in fishing activity in regions or areas eligible under regional Objectives, by including appropriate new criteria for Objectives 2 and 5b.

Source : CEC, 1993a.

of Objective 1 has been increased across the whole of the Community and in the UK, the Highlands and Islands Enterprise area and Merseyside now join Northern Ireland with Objective 1 Status. A substantial part of Structural Fund expenditure will go to cities in Objectives 1 and 2 regions; for example, infrastructure investment for cities in Objective 1 regions and initiatives to support employment training in urban areas suffering from industrial restructuring or Objective 2 regions. New criteria in the amended regulations for Objective 2 regions have meant that urban areas like the Lee Valley and East London have for the first time become eligible for Community support.

The European Social Fund

The Social Fund is the key employment and training initiative of the Commission and it was initiated by the Treaty of Rome in 1957. Most of the ESF assistance goes directly to individual member state governments to assist directly the funding of national training and employment programmes. The Department of Employment works closely with the Commission and local authorities and other bodies like the voluntary sector to ensure that the funds are targeted to the most needy areas. The ESF has a budget of over 4 billion ecu and in 1991 the UK was allocated £400 million which supported approximately 12,500 projects. The Social Fund is being reviewed in view of the new policy orientation of Article 123 of the Treaty of Maastrict. Although the Social Fund will contribute to the five objective areas, Objectives 3 and 4 have been redefined. The main change to Objective 3 is that it now covers the integration of young people and those subject to social exclusion, as well as the long-term unemployed. Objective 4 will contribute towards employment opportunities through vocational training, retraining and other measures (thus preventing unemployment by helping workers to adapt to industrial change). The ESF has been particularly important for urban areas as it performs an important role in influencing the urban labour market through training programmes and initiatives.

Community Initiatives

In 1988 the Regulations (Framework Regulation Act 12 (5)), Co-ordination Regulation Act II (1/2), laid down the principle of Community Initiatives (CIs). The Initiatives give the Commission the opportunity to activate special funds for measures of special interest to the Community. Between 1989 and 1993 just under 10 per cent of Structural Fund commitment appropriations were allocated to the CIs. Member states have criticised the CIs as being unnecessary, small scale and bureaucratic. Although acknowledging these criticisms the Commission has maintained that the CIs do perform a useful role as an instrument of Community policy, enabling measures to be undertaken which go beyond national boundaries. Following the publication of the Green Paper on the Community Initiatives (1994a), the Commission has agreed that the financial framework for the period 1994–99 would total 13.45 billion ecu, which represents 9 per cent of the Structural Fund commitment over this period. Many of the existing CIs have made a valuable contribution towards tackling specific problems in the Community and some initiatives like RENAVAL, to support areas suffering from decline in the shipbuilding industry, and the RECHAR programme which aids former coal and steel communities, have a significant impact on urban areas.

Some member states and many city and city region organisations, with support from the European Parliament, made strong recommendations for expanding the current list of CIs to include the new URBAN initiative which specifically assists urban areas. The URBAN initiative has been allocated 600 million ecu from the Structural Funds with 400 million ecu designated to towns situated in Objective 1 regions. Consideration of eligible areas concentrates on urban neighbourhoods in cities belonging to Objective 1, 2 and 5b regions but with a population of more that 100,000. There are some 350–400 cities in the EU which meet these criteria but only a limited number of cities will be assisted under this initiative (the Commission expects that a maximum of fifty projects will be supported). URBAN will target areas with high levels of unemployment, with a decayed urban fabric, bad housing conditions and a lack of social amenities. Programmes under the initiative need to be integrated, innovative and provide a basis for exchange of experience in the national and European context. Likely programmes to be supported by the initiative include measures relating to economic development, the application of technology, environment improvement, transport, training and education, crime prevention and security, energy efficiency, infrastructure and social facilities.

The European anti-poverty programmes

Between 1975 and 1980 a programme of pilot schemes and studies to combat poverty in European cities was undertaken. In the UK, 15 projects were 50 per cent co-funded by the EC with up to £200,000 being allocated per project. Between 1985 and 1989 a further programme was introduced which focused on particular groups such as the unemployed, the elderly, single parent families and second generation migrants. The third programme, running from 1989 to 1994, with a budget of roughly £39 million, extended and built on the experience gained from the previous two Community programmes. Poverty 3 was more ambitious as it emphasised the role of partnership, participation and a multi-dimensional approach to tackling social exclusion. Poverty 3 finished in June 1994 and in September 1993 the Commission published a proposal for a new 5½ year programme. The budget for a new anti-poverty programme known as Exclusion I will be in the region of 121 million ecu. The Commission's proposal was considered by the member states in April 1994. In Edinburgh, the Pilton Partnership is one of the twenty-nine model projects supported by the Poverty 3 programme. Although the resources have been small (about £2 million over a four-year period), the programme provided a focus for extensive and constructive community participation, which in turn provides a sound basis for the implementation of other projects.

European networks

A major growth area for the Commission has been the support for city networks. The EC contributes financial assistance to European networks between cities within the framework of the RECITE interregional co-operation programme. Assistance is also given to cities to co-operate with counterparts in Eastern and Central Europe through the ECOS Programme, also for the co-operation of cities in the South Mediterranean through the MED-URDS programme. The RECITE programme is open to any local or regional authority with responsibility for a population of more than 50,000. The aim of the RECITE programme is for each network to promote the economic interests and development of its core members, enabling the programme as a whole to strengthen economic and social cohesion within the Community. Examples include the Eurocities network which was established in 1986 and is a network of 'second cities' and includes the cities of Edinburgh and Birmingham. Another is the *Quartiers en Crise* network which began in 1989 as a means to foster exchange and co-operation in towns and cities experimenting with integrated approaches to urban development, and which involves Paisley, Belfast and Manchester.

Urban pilot projects

The Commission has recognised the important role to be played by facilitating the exchange of ideas between cities and regions on how to improve the effectiveness of their urban policies. This has been achieved through the undertaking of pilot projects which explore innovative ways to address urban problems, or as already mentioned the exchange of experience or ideas through the promotion of city networks. Both types of initiative have been funded through Article 10 of the Regulation 4254/88 for implementing the regulations regarding the ERDF which caters for the financing of pilot schemes which encourage co-operation and sharing of experience (RTPI, 1994, p. 91). To this end the Commission has co-financed 21 urban pilot projects designed to test out new ideas in the delivery of urban policy within the Community. Three areas for policy intervention were identified by the Commission and include, firstly, peripheral and inner city housing problem areas, where access to jobs and training excludes many people from economic activity; secondly, projects were supported which would bring together both economic and environmental goals; and finally, the Commission saw the need to facilitate projects which would help to restore economic and commercial life in the historic centres of cities where for various reasons the city fabric has been allowed to decay. European cities which have gained support for projects concerned with economic development in urban areas with social problems include: London, Marseilles, Rotterdam, Brussels, Lyon, Copenhagen, Alborg, Paisley, Groningen and Bremen. Cities involved with environmental actions linked to economic goals have included: Belfast, Athens, Neunkirchen, Madrid, Gibralter and Stoke-on-Trent. Finally, the urban pilot projects have been used to assist in the revitalisation of the historic centres of European cities which includes work undertaken in Lisbon, Thessaloniki, Dublin, Berlin and Genoa.

Europeanisation of Urban Policy?

For some European cities the process of re-urbanisation in the late 1980s meant that urban authorities have to deal with the less desirable aspects of growth, including high land costs, congestion and the degradation of the urban environment (CEC, 1991; 1992a), while other less competitive European cities have continued to be affected by economic decline as a consequence of economic restructuring and the globalisation of the world economy (CEC, 1991; Hall, 1994). Through the mid-1970s and into the 1980s there was growing awareness of the serious economic, social and environmental problems confronting many urban areas across Europe (CEC, 1987; Cheshire and Hay, 1989). This in turn has led many local authorities and national governments to introduce a wide range of local economic development policies aimed at addressing specific issues such as unemployment and combating economic change. The EC, through aspects of European economic and social policy, environmental policy, the anti-poverty programmes and initiatives to combat social exclusion, also attempt to address these issues. The economic position of the EU has advanced through the introduction of the Single European Market (SEM) and more recently has been strengthened by the signing of the Treaty of

Union. Although economic changes have made urban areas important centres not only of economic but also of political power, in the post-Maastrict Europe, a debate is emerging over the need for a more explicit European-wide urban policy. Currently, almost all European cities have some areas or districts which have been run down or suffer from urban decay. Cities and city regions are arguing the case for a stronger led role from the EC in urban affairs and if necessary a further 'Europeanisation' of the policy process as the problems which the Community is trying to address are increasingly being concentrated in urban environments. The challenge for the EU is how to meet the needs of a significant and growing minority of European citizens who have been unable to benefit from the Single Market and have thereby suffered from poverty, unemployment and other forms of economic and social disadvantage.

Many local authorities and organisations involved with local economic development in urban areas recognise this shift towards a 'Europeanisation' of urban policy and initiatives. This Europeanisation of the policy process can be identified through the policies and programmes initiated by the EU which have and will continue to impact on the urban environments. This can be clearly demonstrated through the workings of the Community's Structural Funds, environmental policy, employment training, social policy and the development of trans-European networks of co-operation. Overall the impact of Europe is relatively small in financial terms when compared to the expenditure of individual member states on urban policy and regeneration programmes. But the growing array of European policies and initiatives which impact on urban areas is likely to have far-reaching consequences for the future development, organisation, funding and implementation of urban policy in the UK.

The need for information about the operation of policies and initiatives is a pivotal part of the whole European policy process as the Commission is increasingly concerned about the need for proper monitoring and evaluation. The Commission has published its fourth annual report on the implementation of the reform of the Structural Fund (CEC, 1994c). This document provides an overview assessment of the utilisation of the budget resources of the Structural Funds and particularly lays stress on the assessment work undertaken on the initiative of the Commission or the member states in line with the principles established in the 1988 reform of the Structural Funds. Subsequent to this, the Commission proposed changes to the Structural Fund Regulations which were adopted by the European Council on 20 July 1993. The next programming period runs from 1 January 1994 to 31 December 1999. As part of the implementing procedure the revised regulations make a clear distinction between the three stages of scrutiny required in order to implement the Funds, i.e. appraisal, monitoring and evaluation. After the Edinburgh meeting of the European Council in December 1992 it was agreed that prior appraisal and *ex post* evaluations should be undertaken. The main purpose of this appraisal is to assess whether the Regional Plan for a particular Objective region represents good value for money in terms of medium-term economic and social goals. Each programme region also comes under the scrutiny of individual monitoring committees which are made up of representatives from the regional and national level and representatives from the Commission. Evaluations are also required to assess the overall impact of the Community programme on the local economy and these evaluations should also include a measure of the progress towards economic and social convergence. Similarly assessments have to be made concerning the effectiveness of measures in stimulating economic development and commenting on the value for money achieved from the resources deployed.

In the case of the Strathclyde Integrated Development Operation (IDO) the Scottish Office funded an independent consultant to carry out an interim assessment of the IDO in 1992 (Scottish Office, 1993). The Strathclyde IDO was approved by the EC in December 1988. The project was to involve the co-ordinated use of European and national resources in the regeneration of the Objective 2 area of Strathclyde. Objective 2 involves converting the regions, frontier regions or parts of regions (including employment areas and urban communities) which are seriously affected by industrial decline. This area covered the whole of Strathclyde region, while excluding Argyle and Bute and Kyle and Carrick plus the northern part of Dumfries and Galloway Region (Scottish Office, 1993). The lessons learnt from this interim assessment were incorporated into the monitoring and evaluation strategy for the West of Scotland Regional Plan 1994–99 which has been submitted to the Commission for approval.

Unlike many aspects of UK urban policy the EC is extremely rigorous over programme and policy evaluations. Sometimes the Commission has been criticised for being too concerned with the *procedures* for monitoring and evaluation. Substantial elements of programming budgets for the Community Initiatives and the anti-poverty programme have been used for such purposes and subsequently these resources have not been utilised for individual projects within the programme. Policy evaluations undertaken by the Commission have therefore tended to concentrate on individual programmes or on the operation of the Structural Funds within the Objective regions. If the Commission is serious about addressing urban problems, and in keeping with the notion of subsidiarity, further research and evaluations should be undertaken to assess the overall impact of European policies and programmes in the wider context of local and national based strategies.

Conclusions

During the past fifteen years there has been a slow but continual withdrawal of UK central government expenditure on urban policy. As the problems experienced by urban areas continue UK urban policy has been characterised by the introduction of a succession of different policy instruments such as the Enterprise Zones experiment, the Urban Development Corporations, City Challenge and the recent introduction of the Single Regeneration Budget. As the UK national government introduces new mechanisms and policies designed to deliver efficient and cost effective policy, many financially depleted urban local authorities rely on Europe as their main source of funding. Although the EU has no mandate as yet for the creation of a European-wide urban policy it is clear that the activities and policies of the EC do in fact either indirectly or directly impact on urban areas. As concern for the economic, environmental and social development of urban areas continues the trend towards the Europeanisation of urban policy in the UK is a distinct possibility. As the European influence over the urban environment increases, calls are made for more comprehensive evaluations of the impacts of urban policy on UK cities. Policy-makers and researchers must, therefore, begin to appreciate the contribution and significance of the key European policy instruments, the Structural Funds, the Community Initiatives, environmental policy, education and training, and policies geared towards tackling the multi-dimensional aspects of social exclusion, as the resolve of the Commission to improve the economic, environmental and social conditions in urban areas across the Community strengthens.

PART TWO

Methodology and Dilemmas
in Urban Policy Evaluation

6

MEASUREMENT, UNDERSTANDING AND PERCEPTION: THE CONFLICTING 'REALITIES'

Peter Meyer

Introduction

Economic evaluation of urban policy has come to be demanded more and more by policy-makers at both local and national levels in both Europe and America. Private sector intervenors in the policy process also call for (and offer) quantitative appraisals of alternatives. Most commonly, the claims to scientific credibility of such assessments are based on the sophistication of the analysis of returns on public sector investments. Return-on-investment calculations, in turn, require consistent measurement of impact in social welfare terms and their translation into monetary terms (Gramlich, 1981; Mishan, 1976; Sugden and Williams, 1978; Thompson, 1980).

The political viability of any evaluations, and their credibility and influence in shaping policy, rests, however, on their acceptance by a broad political public. However sophisticated a quantitative analysis, its real value as a resource for decision-making lies in the extent to which the assumptions underlying the measures employed are accepted as consistent with public beliefs, perceptions and understandings. The problem of the acceptability of evaluative measures is compounded by the fact that different urban policy constituencies may perceive different 'realities': they may disagree in their perceptions and beliefs and/or may fail to share a common population or geographic frame of reference.

This difficulty is starkly evident in conflicts between the local and national states in Britain and the United States. Both nations have had policies held dear by local officials that were abolished by the national state and have experienced imposition of national programmes on municipalities that resisted them. Examples of the former include the abolition of the US Community Development Block Grant and Urban Development Action Grant programmes and government constraints on the economic development powers of the local authorities in Britain. Creation and imposition of Urban Development Corporations and Training and Enterprise Councils in the UK and an array of externally mandated placements of locally undesirable land uses (LULUs) on municipalities and neighbourhoods in the USA exemplify the latter.

Divergence in world-view and in valuation of outcomes across different geographic frames of reference is, perhaps, obvious. The same differences exist, however, with respect to non-spatially-defined groups of policy constituents. 'Objective' policy and maximisation of any collectively defined or accepted welfare function may be unattainable in a socially diverse environment. The greater the relevant population diversity, the weaker will be any comprehensive evaluation, while, at the same time, the stronger may be the political imperatives for arriving at some summary measure (Cole, 1993). Inevitably, then, the higher will be the tendency to ignore or run roughshod over minority or unpopular perspectives, unless the political (and economic) power of their proponents are exceptionally great (Bromley, 1991; Gould, 1981). If, however, promotion of democracy and community are to have any role in public policy objectives, such imposition of a 'majority will' is inappropriate.

Recent increases in ethnic and cultural diversity in cities across the globe and new and increasingly complex class divisions worldwide have exacerbated the evaluation problem. These changes appear to have greatly accentuated the incompatibility of different 'realities'. For our purposes here, it matters not whether these divisions constitute fractures of the type that Jacobs (1992) argues were wrought by the Thatcher and Reagan regimes or are the result of social ecological processes rather than conscious political acts, as Dendrinos (1992) has suggested may be the case.

Yet another source of difference in understandings of urban processes and policy outcomes lies in the world-view of analysts themselves. These differences need not even be ideological; they can reflect disciplinary biases and little more. Thus, for example, the economic analyses that ignore the presence of conflict and power differentials may have little room for consideration of distributional issues that are central to the concerns of social welfare analysts, and are likely to figure more prominently in the calculations of sociologists and political scientists than in those of economists. The analytical assumptions, processes, and thus the findings of urban planners may well be totally inconsistent with those of analysts guided by a belief in the superiority of market processes, even when all operate in the context of a predominantly private property and private capital based market system.

This difference, while of substantial interest to theoreticians (who, them-selves, may exemplify the conflicts), can be disregarded here. Whatever the epistemological differences that may distinguish alternative analyses, in any given decision-making and judicial review context, some dominant paradigm will prevail, and all analysts seeking to influence policy decision outcomes will be forced to present their arguments within that dominant frame of reference, whatever the origins of their perspectives and findings. Moreover, while it may be argued that community members may suffer distorted welfare functions ('false consciousness') as the result of private sector publicity efforts, the well-being impacts based on those valuations are real, and have to be accepted as the appropriate guiding principles so long as individuals' values hold constant.

The objective of this chapter, then, is to derive preliminary guiding principles for evaluation efforts in the current context of great diversity within and between urban areas and conurbations. I begin by reviewing the requirements of evaluation and the conflicts between analytical method and political significance. Next, I consider the objectives of evaluations and the conditions that must be met to assure any degree of scientific accuracy in such efforts. Third, I turn to examples of policy evaluation problems. I conclude by turning to the methods and ethics I consider appropriate to evaluative efforts in diverse and complex urban settings and offering a decision-making structure that might provide appropriate outcomes.

The Requirements for Consistent Evaluation

Any evaluation is inherently *comparative*. Even if only one possible action is considered for a 'go/no-go' decision, a comparison is required – in this case between the expected outcome with and without the programme or policy. There is thus a need for a consistent measure of the volume of each of the relevant outcomes and a means for assigning a value to each unit of each impact generated. This requirement generates two analytical problems that, in practice, present a political, rather than scientific, face.

First, there is the issue of which impacts are, in fact, relevant. Different constituencies will disagree on the question of which effects are to be counted. In some instances, the disagreement will be over what the true objective of the programme or policy is – a matter of political perception. In others, perhaps more intractable, the divergence in perception may lie in what the ancillary or unintended effects of a policy actually are. Here there may be disagreement about the probable counter-factual – the conditions that would have prevailed in the absence of a particular intervention – and the focus is typically on unintended negative consequences or externalities.

Second, even if there were agreement on the impacts to consider, alternative public actions may produce different *mixes* of those impacts – greater additions to incomes in one case, but combined with greater health risks, perhaps, than for an alternative. Consistent *numeraires* for valuing the different combinations are required for comparison of the mixes emerging from different interventions. Even if not all impacts are translated into monetary terms, health risk tradeoffs with income – or with aesthetic effects or distributional impacts, and so on – have to be measured the same way for all alternatives. Some minimum of commensurability between the impacts considered is essential to this process. A policy that guarantees a certain number of deaths in return for income increases thus may be impossible to evaluate.

In the context of any degree of diversity in a society, these two problems appear to doom any claims to universal objectivity: even if the volume of each impact is agreed upon by all parties, the valuation of those outcome units is not likely to be consistent across all constituencies. Local officials, for example, can be expected to disregard non-local impacts, though these may be critical in the calculations of national officials. An income increment, a health status effect or a redistribution across population sub-groups is not likely to be equally valued by the rich and the poor, the white and the non-white, or even the residents of different neighbourhoods within a city, however strong their socio-economic status commonalities.

Even in the absence of ethnic, religious, racial or spatially defined forms of diversity, the variation that neo-classical economists acknowledge to be present across individuals' 'utility functions' will result in divergent valuations of the impacts. The valuations *cannot* be aggregated consistently into a common social preference statement, as Arrow demonstrated in 1951. Inevitably, the assumptions made to simplify reality in an evaluation model will not be equally acceptable to all parties. Some localised, neighbourhood-specific or ethnically-specific values and beliefs are bound to be overlooked or not well served by any evaluation construct. (In effect, if we acknowledge the existence of *any* conflict over resources in a society, the very concept of a universal 'social welfare function' is inherently chimeric.)

Objectivity is thus a logical impossibility. Any evaluation reflects some limited subset of social values. Therefore, an evaluation model should be held, at a minimum, to a simple standard: does it accurately represent the policy objectives inherent in the legislation or regulations whose implementation is being evaluated?

Ironically, in analysts' extensive efforts to arrive at the non-attainable standard of objectivity, many evaluative constructs fail this test. Excess attention may be given to legitimation, either through strengthening the mathematical consistency of measures of output or through addressing the outcomes of greatest importance to the constituencies affected. The intent of the intervention can thus be overlooked. (At times, the policy may be *designed* to do damage to a particular constituency, so an apparently negative impact is actually positive in terms of the new programme's objectives. The intent of the policy should rule the valuation decision in this instance – even though such a procedure is clearly not 'objective' in the absolute sense, and may contribute to significant welfare losses; the most obvious example of such a policy is the 'principle' of progressive taxation, designed to 'damage' the richer in order to permit redistribution on some level to the poorer through government programmes.) The political acceptability of a policy objective may be at issue, but the evaluation should still serve the original legislative or regulatory intent, until that intent is modified.

The Objectives of Evaluation – and Conditions for Accuracy

A good evaluation should satisfy the three Es of Effectiveness, Equity and Efficiency. Whatever the programme, an evaluation should permit it to be improved in these three dimensions, however they are defined by the policy-makers. (Note, however, that policy-makers' definitions may either be shaped by, or may be in direct conflict with, the definitions of one or more of the constituents of an urban policy.) In order for these objectives to be served, however, the evaluation itself must meet minimum criteria of political acceptability. That is, if the evaluation is not credible, it will not positively affect the political process that perpetuates, modifies or terminates programmes, regardless of its content and relevance to programme objectives.

The evaluation should promote *Effectiveness* by providing measures of the extent to which the objectives of a policy are met by a given programme. Thus, for an evaluation to be able to assess the extent to which a programme achieves its ends, the policy objectives it is designed to serve must not only be well known and understood, but they also must be broadly accepted as legitimate by the constituencies affected by the policy and the programme.

Equity, whether we like it or not, is defined situationally; it is not synonymous with equality. Some programmes may increase inequality in order to serve equity objectives. (Compensatory programmes such as employment training in minority communities may lead other low income constituents to challenge the inequality of resource allocation. If the programme objective is raising minority employment levels to those of whites in an area, equity in this outcome measure may call for inequality in distribution of programme resources.) Conflict between different groups (and, for that matter, between individuals within groups) over how equity is to be defined is, effectively, inevitable.

The third objective, that of promoting *Efficiency*, is the intent that requires a comparative element in all evaluations. It is not enough to say that a given programme or policy is effective in attaining its objectives. A good evaluation will ask about the costs (monetary and otherwise) of the means used to attain the intended end. The evaluation should ask, 'Can the same effect be attained at lower monetary, social and environmental costs?' Only if the answer is in the negative can the policy in question be considered to be genuinely efficient. (Note that this definition of efficiency goes well beyond the narrow perspective of 'value for money' – which it should.)

A distinction between programmatic *activity* and *impact* must be drawn if an evaluation is to promote the three Es. All interventions have specific *outputs*: numbers of people trained, buildings razed or rehabilitated, contacts with clients made, and so on. If, however, the trainees find no jobs, the cleared land or new buildings find no buyers, or the client contacts result in no further interaction, these activities may have zero or even negative *outcomes*. The *output* measures reflect activity, rather than impact. An evaluation must have measures of (the policy-intended and inadvertent) *outcomes* that are important to the programme's multiple constituents if it is to have any chance of increasing effectiveness, equity and efficiency.

Empirical Evidence and Evaluation Experiences

Any set of cases for discussion of the problems of evaluation must be chosen arbitrarily, given the range of relevant concerns. I consider first an egregious case of evaluation based on inappropriate criteria, then turn to a group of brief US and UK cases that, taken together, suggest a direction for action to improve evaluation practice, and conclude with a description of an adversarial process that arrives at flexible co-mingling of objectives and valuations in a US regulatory context.

The use of inappropriate criteria

Poverty eradication efforts in the USA have long been governed by a concern for 'vertical efficiency'. The term refers to the proportion of funds going to a given 'target population', in this case people below the poverty line. Opposition to the non-deserving 'getting something for nothing' necessitated this criterion. Its net effects have been findings of programme failure whenever funds flowing to low income households began to lift them out of poverty (since the recipients moved above the poverty line and vertical efficiency declined). No poverty reduction effort could survive this case of application of a politically imperative but programmatically irrational and inappropriate criterion. Policy objectives needed to be explicitly addressed and made explicit.

Poverty policy in a single country raises another evaluation problem: that of impacts on jurisdictions or peoples that are not party to policy decision-making.

Was the placement of a major nuclear power plant at the tip of Cap de la Hague a good idea? The French thought so – but prevailing winds suggest Britain would be hurt by an accident, not France. Is job generation in one country that impoverishes another good social policy? Obviously, in the absence of a trans-national public policy decision-making apparatus, such questions will never be addressed by those with the power to implement decisions. Nonetheless, there is a real question of the appropriateness of the *geographic frame of reference* of evaluative measures. (Note, however, that this discourse by implication suggests the legitimacy of some role for the national state in constraining economic development efforts by individual localities within its borders.)

Some examples

Neighbourhood development proposals (whether old Model Cities efforts in the USA, the old Industrial and Commercial Improvement Areas in the UK, or the transformations proposed by UK Urban Development Corporations) run into conflicts between the interests of current residents or occupants and those of potential in-migrant households or businesses. No amount of community participation by current occupants will produce objectives (or outcome valuations) that could serve rehabilitation or population change intents. A choice of whose interests should be served is thus unavoidable. (The fact that ideology may cloud the perceptions of choices by those with the power to decide and lead to irrational objective specifications, as evidenced by the various UDC fiascos (Imrie and Thomas, 1993), does not lead automatically to the conclusion that the objectives of current neighbourhood occupants should always take precedence: the choice problem remains.)

A focus on neighbourhoods raises another policy planning and evaluation dilemma: the inability to argue by analogy. Valuations of outcomes are always difficult to estimate, and it is common practice to borrow from findings of prior evaluations at other sites or of other projects. The socio-economic specificity and uniqueness of *countries* is admitted by the Organisation for Economic Co-operation and Development (1991). If national environmental policies should be tailored to the differences between OECD members, with common developed market economies, then how unique must be *neighbourhood* development efforts – and how inappropriate any arguments by analogy? I would suggest strongly that neighbourhood and community concerns, since they are more likely to represent *some* expression of a social welfare function than any objective promulgated by the national (or municipal) state, should shape public policy interventions. Thus, in the ideal, community valuations of outcomes should be the measures used in evaluations of external interventions in localities. Unfortunately, this democratic ideal is rarely attained.

Reclamation of abandoned and polluted land, especially in older industrial areas, may be essential to the economic revitalisation of urban cores. Any decision on whether or not to rehabilitate a given property must address the divisions and different perspectives inherent in urban programme evaluation, including the issue of neighbourhood specificity. Consider the variety of possible parties to such a planning and permitting process and the inevitability of conflict between their perspectives on essential but variable aspects of a reclamation policy: (a) who pays for clean-up, (b) for what future purpose is the land to be used, (c) how extensive should the clean-up effort be, and (d) who is to bear the responsibility for inadequate reclamation efforts (Yount and Meyer, 1994).

The parties include: (1) current private sector owners of abandoned or redundant land and buildings; (2) local government that has an interest in generating productive use of the land (and thus taxes on new property values, at least in the USA); (3) the regional or national state that sets environmental reclamation requirements; (4) property developers, concerned with profit, with potential consumer attitudes towards previously polluted lands, and thus with the stigmatisation that could result from publicised clean-up efforts; (5) financiers, who are only concerned about their returns, not developers' profits, but who may have fears about pollution liability if they have to foreclose on bad loans; (6) neighbourhood residents, whose primary concern may be for removal of potentially dangerous pollutants, not future land uses; and (7) other urban residents whose property values may fall due to new lands becoming available. The list could go on – but which parties constitute the 'community' whose interests I have suggested should prevail? How should their divergent interests be arrayed, and weighed?

Choice of technology in waste reduction, treatment and disposal raises all the conflicts inherent in any land-use conflict, but adds the spectre of all the environment–employment trade-offs that have dogged emerging ecologically sensitive technologies (Meyer, 1994). This choice can be seen to be a two-stage process, with technology and waste reduction emphasis decisions coming first, and decisions on waste treatment and disposal following. With respect to technological choices, class, current income, employment status, culture and religion will all influence relative weights. Recognition that environment–employment choices need *not* be a zero-sum game may be a function of education – and of ideology. Age and family status may be major determinants of time horizons in technology choices. The less educated may fear advancing technologies that could cost them their jobs more than would the technologically trained. Class and income source may be major determinants of the valuation of alternatives in imposing the financial burdens of technological change on the private or public sectors.

Waste storage, treatment and disposal, unlike production technologies, are activities located at specific points in space. Thus they raise all the local/national goal and valuation conflicts already highlighted. The stark local-cost/national-benefit character of these decisions highlights the potential of monetary transfers as means for resolving conflicts. The problem of immediacy of financial need remains, however, as poor residents will be willing to permit local disposal at a lower compensation cost than the more affluent, and will be less inclined to recognise the damage that long-term stigma effects may have on the economic well-being of their children, even if all actual health risks are eliminated. (Note this means that community control may be meaningless if income inequality is sufficiently acute.)

Multi-party decision-making in US utility regulation

The regulatory process by which public utility pricing and provision policies are determined in the United States may provide a guide to a policy evaluation model to recognise and balance multiple perspectives. The example provided here is based on the author's own experience working in the context of the Pennsylvania Public Utility Commission rate and regulations determination processes; similar structures for arriving at policies and practices exist in the other states of the USA.

The parties involved in the determination of the rate structure and payment requirements of an electric or gas company are many, including the company itself (motivated to maximise its profits as a private shareholder-owned firm); and a number of different groups, each provided with different power rates, such as: industrial users (sometimes differentiated by sector), commercial users, local governments (who may testify both on their own behalf as users and on the behalf of their residents or local businesses), schools, households; and special constituencies (the poor, residents of public housing, people with special medical needs for high usage, and the like).

The regulatory process is quasi-judicial, with all interested parties provided the opportunity not only to present testimony, typically offered in written form in advance, but also to cross-examine all others' witnesses and to file rejoinders to cross-examination. A designated 'hearing officer', generally a legally trained person, authorised to take testimony by a politically independent utility commission, presides over the presentation of argument and summarises positions, disputes, bases of disagreements for the commission, to whom he or she also offers a recommendation. In the more progressive states and other jurisdictions, provision is made to assure that those parties to argument over utility charges that have fewer financial resources with which to hire expertise are provided with funds to do so. The hearings process involves confrontation between representatives of a range of divergent perspectives that may agree on some aspects of proposed policy but disagree on others, and tend to exhibit shifting alliances as each issue is raised before the hearing officer. Utility rate issues are thus examined from a broader set of different perspectives than is generally the case for evaluation of the impacts of any other social policy issues in the United States.

Prescriptive Responses to the Impossibility of Objectivity

SUBJECTIVITY RULES! – OK!

This should be the graffito of the evaluator. From this credo, which is no more than the acceptance of reality, emerge principles for more principled evaluations. The evaluation ideal pursues both universality and consistency in valuation of outcomes. In practice, both are jeopardised.

Universality, as we have seen, is unattainable if we concern ourselves with the valuation of all outcomes. However, we may move forward by recognising that there may be universality in the *count* of the volume of outcomes, but not their *valuations*. A parochial local interest may treat non-local impacts as irrelevant, giving them zero value, but not deny that such outcomes result. If the variety of outcomes considered is sufficiently broad and outcome volumes are disaggregated for socially relevant population sub-groups, then partial universality becomes attainable. We are left with specificity and conflict in the valuation of the outcomes, and of different distributions of those outcomes across different population groups. This is not an insignificant problem, but we have narrowed the scope of analytical uncertainty.

Consistency, the treatment of all alternatives in the same analytical manner, depends not so much on the agreement of different groups on how to value outcomes, but rather on the stability of the preferences and valuations of alternatives held by those groups. We can recognise the differences in valuation associated with location, class, education, ethnicity, age, gender and so on and still be consistent: what is required is that those values be applied equally to the outcomes from each alternative.

No project or policy can be accurately ('objectively') represented by a single summary measure, given the multiplicity of constituencies. Yet the comparison of valuations by a number of different constituent groups can be useful for decision-makers. Consider the following three cases of comparisons on a proposed policy: (1) All constituents find the benefits exceed the costs – there is no conflict and the policy goes forward; (2) All constituents find the costs exceed the benefits – there is no conflict, and the policy is rejected; and (3) Some constituents find positive net benefits while others perceive net costs – conflict is present.

This last case may be the norm, but the very process of comparison helps establish the extent of conflict and the ease with which it might be resolved. If the 'losers' are small in number and/or suffer minor losses, appropriate compensation might turn them into 'winners' while leaving all others as net gainers. (Note that the most difficult part of implementing a compensation scheme – identifying the losers – is already accomplished by an evaluation system that tracks different population subgroups.) A new alternative, one incorporating the compensation, may then be added to the choices available.

Such a response is narrowly economistic, however. It is possible to consider whether the losses experienced may be alleviated in other manners. For example, the losses experienced by forcing low income housing (and thus the possibility of racial diversity) on lily-white neighbourhoods may be associated with the area's racism. In such a case, the way to correct the problem may be to add a programme to address the racism itself as part of the new alternative, rather than simply to provide monetary compensation.

When losses and gains are more evenly distributed across a population, the social choice of whose interests have higher weight must be made. Weighting decisions inevitably involve damage to some for a (presumably greater) benefit to others and to the totality. The use of an evaluation scheme that explicitly distinguishes the impacts on all groups does not solve the choice problem, but it forces it into the open (see Cole, 1993, for an excellent example). This, in itself, is a positive outcome, since concentrated large gains for the rich and powerful combined with minor gains for a majority of the population could be identified as the inequality-generating actions that they are.

The Multi-Party Decision-Making Model Revisited

To the extent that the exposure just discussed is coupled to another procedure, a variant of the explicit adversarial representation of the different interest groups made possible in the US utility regulatory process (whose hearings are made public and accessible to all), then more egalitarian outcomes may even result. Current practice has business interests, the rich and the powerful represented by cost-benefit professionals while other groups scramble to present their concerns as best they can. Acceptance of the inevitability of conflict and the institutionalisation of a 'cost-benefit calculation court' in which all parties may be represented by professional economics 'counsel' would promote conscious and overt, not unintended, uninformed, and easily subverted, decisions on the inevitable conflicts in a diverse and complex society. In effect, the 'hearing officer' would be responsible for arriving at a recommended cost-benefit analysis finding, providing his or her rationale for that finding for scrutiny by the public officials actually responsible for making a decision on a policy or programme.

Such an approach would require that all interested parties be provided with resources to use in presenting their arguments in the hearings process and in challenging others' arguments. The costs involved in such an endeavour would be substantial, and the time delays associated with it could also contribute to rising expenses for many projects with public impacts. The financial burdens of such a procedure would thus be substantial, but the social costs of excluding participation and insight from affected parties could, conceivably, be far greater.

However, the financial burdens are only a small portion of the problems that efforts to implement such a quasi-court would have to overcome. Others include the issue of who would be represented, since some degree of legitimacy for an interested group is needed for it to garner participation resources. (While anyone who wants to can 'intervene' in a utility case in the USA, only some groups – in some jurisdictions – are provided with financial aid to help make their case.) Moreover, splintering of interest groups may reduce their impact, especially if they get resources proportional to their numbers. Thus minority interests may still be under-represented, or amalgamated into a policy position taken by a larger body. (Groups representing minorities, the poor and the elderly/disabled tend to join together to take a common position on utility company rates and regulation issues in US hearings, in order to combine resources to strengthen their voices.)

A depoliticised, open, confrontational and accessible process for argument about cost-benefit valuation will not be a panacea. Many will be excluded if they do not get resources with which to participate, a bias towards one analytical frame of reference may undermine arguments based on another paradigm offered by others, and the very length and cost of the process may tend to delegitimate it in the context of intra-locational competition for projects contributing to so-called 'economic development'. Still, an effort to provide a common setting in which the issues addressed here can be raised and considered by a wide array of interested parties could provide a first step to a more efficient process of multi-party evaluation of urban policies and programmes.

7

THE EVALUATION OF URBAN POLICY PROJECT APPRAISAL

Steve Martin and Graham Pearce

Introduction

The proliferation of central and local government programmes aimed at promoting the economic and social regeneration of inner city areas in the 1980s was accompanied by a series of *ex post* evaluations of the outcomes of these initiatives. As a consequence, considerable progress was made in the development of methodologies which provided more accurate impact assessments and established 'value for money' criteria by which the performance of both individual projects and programmes could be judged (Bovaird, Gregory and Martin, 1991). Researchers placed particular emphasis upon the relationship between anticipated outcomes and policy objectives and on the efficiency of the procedures for project and programme management. By contrast, they paid comparatively limited attention to methods of *ex ante* appraisal. This is surprising since appraisal determines which projects are financed from the public purse and is, therefore, a crucial stage in the policy-making process.

The primary purpose of appraisal is to evaluate a project's potential to meet its objectives and, thereby, contribute to wider policy objectives. The appraisal process should, therefore, ideally draw together a wide range of information about a project's likely outcomes. Two main issues need to be addressed. Firstly, will the project, as formulated, meet its own objectives and the wider needs of the locality? Secondly, how does it compare with other projects which have similar objectives and are in competition with it for resources?

Given the desire on the part of all governments to secure the best possible 'value for money' from their spending on regeneration programmes the accurate, *ex ante* assessment of the relative costs and anticipated benefits of regeneration projects is likely to become an increasingly important feature of resource allocation decisions. Indeed, this trend is clearly evident in the new proposals for the allocation of European Structural Funds which attach much greater weight to project appraisal than has been the case hitherto. The systematic identification of the *opportunity* costs of supporting particular projects or programmes is also likely to become increasingly important in the UK given the emphasis placed on competition between projects and areas for urban regeneration funds by recent initiatives such as City Challenge, Single Regeneration Budgets (SRBs) and Regional Challenge.

This chapter examines these issues in the context of the key central government urban policy measures operating in the UK in the late 1980s – the Urban Programme (UP), City Grant (CG), Derelict Land Grant (DLG) and the Urban Development Corporations (UDCs).[1]

The Meaning of Project Appraisal

Appraisal is sometimes considered to be synonymous with 'assessment' or 'evaluation'. However, applied to the planning and assessment of investment projects it can be defined more narrowly as an exclusively *ex ante* procedure aimed at determining a project's objectives, examining options and weighing up its costs and benefits before deciding whether to proceed with it (HM Treasury, 1988). Project appraisal in this context is, therefore, a procedure which is designed to select the 'best' projects and includes the identification and 'sieving' of a wide range of alternative schemes before subjecting them to formal appraisal. The value of 'formal' appraisal is, therefore, heavily dependent upon the rigour with which alternatives have been identified and investigated at an early stage in the process.

In practice, however, there is a tendency in government to think of project appraisal as a much narrower exercise, involving the 'checking' of a fairly limited pool of potential actions against a narrow range of criteria. Moreover, appraisal, whether in its broad or narrow sense, is not normally geared solely, or even mainly, to the identification of those projects which provide good 'value for money' and most systems are influenced, to a greater or lesser degree, by other considerations. Appraisal systems are normally required to fulfil a number of functions. They must not only incorporate 'rational' procedures which enable those projects which are most likely to meet programme objectives to be identified, but must also demonstrate that the formal administrative procedures have been followed in a way which ensures accountability for public expenditure. Moreover, appraisers will often be under considerable pressure to ensure that sufficient projects are approved to satisfy a political imperative to achieve spending targets.

Alternative perceptions regarding the role of appraisal are important influences upon the weight attached to different aspects of the appraisal process. If 'good practice' is seen as ensuring that the 'best' projects are selected, the use of rigorous and relevant appraisal criteria becomes paramount. However, if it is visualised primarily as a means of demonstrating that appraisers are publicly accountable, the emphasis will be upon the verification of a project's eligibility, the careful documentation of the reasons for decisions and the provision of accurate management information. Alternatively, if action on the ground is the overriding consideration, projects will need to 'pass through' appraisal systems as fast as possible and there will be a premium on the minimisation of 'bureaucratic' delays and obstacles.

Pressures to secure spend and/or satisfy demands for accountability shifted

the emphasis of appraisals of urban projects in the 1980s away from the identification of those which were likely to make the greatest contribution to urban
regeneration. Perhaps the best examples were the UDCs established in London
Docklands and Merseyside in the early 1980s. These urban policy 'flagships'
were intended to use public sector funds primarily to create momentum in
the local property market and to establish optimism and commitment from
commercial interests. Borrowing heavily from US experience, this approach
was in marked contrast to the approaches to urban renewal adopted by UK
governments in the 1960s and 1970s, being characterised by an emphasis which
was explicitly 'anti long-term strategic planning, anti almost any published plan
at all, freewheeling, free booting . . . concerned only to exploit opportunities
as they arose' (Hall, 1988). In these circumstances policy might be seen as
having been driven as much by a desire to achieve tangible outcomes within
the relatively short life-span of the UDCs as by the need to ensure that they
would enhance the long-term prospects of their areas.

The commercial orientation of many UK urban policy initiatives in the 1980s
was also reflected in the way in which the enhancement of land values and
the leverage of the maximum level of private sector investment were used as
surrogate measures for regeneration activity. The Urban Development Grant
(UDG) programme, introduced in 1983, was based directly upon the experience
of the Urban Development Action Grant programme in the USA and reflected a
concern to attract private sector investment into inner city areas. The so-called
'gearing ratio' (the relationship between private and public sector funding)
therefore became a dominant performance measure and appraisal criterion.
Had a different policy stance been taken by central government quite different
appraisal criteria might have been adopted, leading to the selection of a range
of alternative projects with quite different regeneration outcomes.

A Model of the Appraisal Process

In essence appraisal systems for inner city projects need to fulfil three basic
functions:

- determining whether projects match the eligibility criteria spelled out in
  government guidelines;
- identifying the appraisal procedures that are most appropriate for each
  project;
- assembling relevant, comprehensive and accurate information to enable
  judgements to be made about the likely costs and outcomes of alternative
  projects.

Furthermore, appraisal systems can be usefully considered as consisting of a
number of stages and dimensions which address fundamental questions. These
include the:

- identification of a project's objectives, including its relationship to other
  projects and broader programme objectives;
- generation of alternative options relating, for example, to the form, quality
  and timing of a project;

- identification of the benefits and dis-benefits associated with a project;
- identification of the interrelationship between the costs of a project and its value on completion;
- testing of a project's feasibility, assessing, for example, its technical practicability, the availability of finance, its viability with reference to future market conditions and the level of risk it involves.

Whilst all appraisal systems need to incorporate these broad elements, the detailed procedures associated with individual inner city programmes should be tailored to their specific objectives. All of the major urban policy measures of the 1980s shared the objective of contributing to the regeneration of inner city areas but each had a different emphasis. They varied markedly in terms of their size, the characteristics of the projects and the organisational framework within which they operated. Prior to its demise in the early 1990s, the Urban Programme operated in nearly 60 local authority areas and funded some 12,000 (mostly small-scale) projects at any one time. By contrast fewer than 100 City Grants were approved each year but, like many projects supported by DLG and the UDCs, they involved major land reclamation, infrastructure provision and new building developments, which attracted large-scale public and private sector investment.

The locus of project appraisal also varied between the four programmes. By virtue of their substantial delegated powers, UDCs retained considerable discretion over project appraisal, with all but the very largest schemes being approved by their Boards rather than Department of the Environment (DoE) staff. The DoE's Regional Offices played a far more pro-active role with regard to both DLG and the Urban Programme, not only influencing the direction of the programmes, but also appraising and scrutinising individual projects. However, the degree of involvement varied considerably between UP and DLG projects, depending upon the characteristics of the projects, the priorities and practices of individual Regional Offices and the past performance and the perceived competence of individual local authorities. By contrast all applications for City Grant were assessed by a specialist team located within the DoE's headquarters in London.

Key Weaknesses in the Appraisal Process

Each of the four major inner city programmes had particular, individual weaknesses. However, two generic sorts of problem were common to all four. Firstly, a number of 'technical issues' were generally dealt with inadequately by appraisers – thus casting doubts upon the accuracy of the judgements which they made about the scale of project costs and benefits. Secondly, there was a fairly widespread failure to appraise projects within a wider policy context, leading to the approval of what were often, in effect, lists of individual projects rather than coherent programmes of co-ordinated actions (Martin and Pearce, 1993). As a result, even if appraisers produced accurate assessments of the likely outcomes of individual projects, they often failed to examine the contribution that these would make to the achievement of overall policy objectives.

Technical Weaknesses

Impact measures

In pursuing the primary objective of appraisal identified above (p. 101), the principal task of appraisers is to identify the benefits, dis-benefits and costs of projects. This implies the need for accurate measures of the inputs, outputs and impacts of proposed projects in order to determine their 'value for money' in terms of the '3 Es' economy, efficiency and effectiveness. It is clear, however, that in practice many appraisers gave undue emphasis to the costs of projects and neglected outcomes. This is perhaps understandable given that the level of inputs is usually easier to estimate than the scale of benefits which will result from projects. However, it meant that a great deal of attention was focused on the issue of economy and much less importance was attached to effectiveness.

Moreover, assessments of potential outcomes often focused upon immediate and intermediate outputs (such as the number of training places to be provided, area of land to be reclaimed, the length of road which would be constructed or the number of workshops to be built). As a result appraisal systems generally failed to give sufficient attention to 'higher level' (or 'final') outcomes (for example job generation, increases in aggregate income, improvements in trainees' employment prospects and enhanced productivity of businesses). Whilst this is understandable given the considerable problems involved in estimating final outcomes, it inevitably meant that the strength of the link between project outcomes and the achievement of programme objectives remained uncertain. This limitation was also reflected in central government's approach to the monitoring and evaluation of inner city projects which continued to rely upon easily quantified, immediate outputs as surrogate measures of final outcomes.

Many appraisals also lacked an adequate longitudinal perspective. The likely longevity of project benefits was rarely assessed and there was a risk that projects which yielded substantial immediate outputs would be selected in preference to those which yielded greater long-term benefits. An example was the way in which appraisers were slow to adopt the notion of 'full-time equivalent job years' and sometimes failed even to differentiate between 'permanent' and 'temporary' (for example, construction) jobs.

Additionality and equity

Most appraisal systems neglected what have been referred to as the fourth and fifth 'Es', additionality and equity (Gregory and Martin, 1988). Additionality has two dimensions. The first is based upon an assessment of the minimum level of public sector funding required for a project to proceed. The second refers to the extra output or the 'value' added which will result from a completed project. In practice, however, these two dimensions are closely related since whilst a project may have the potential to produce substantial outputs (for example, in terms of additional private sector investment or jobs), if it will proceed in the absence of public assistance, these 'additional' outputs should be discounted.

The assessment of additionality is therefore a demanding task which requires appraisers to judge what is likely to happen if public monies are not given to a project.

This issue has received increasing attention from researchers (for example, Martin, 1989; Storey, 1990; Foley, 1992) but conceptual developments have not been translated consistently into improved appraisal practices. There were significant variations between inner city programmes in terms of the rigour with which additionality was assessed, with many appraisers simply assuming that a project would not go ahead without public subsidy and that all of the anticipated project outcomes could therefore be attributed to it. In so doing they overlooked the possibility that a project might not in fact require public assistance or could merely duplicate activities which would have occurred in its absence. The latter oversight was symptomatic of the lack of attention to the potential negative impacts of projects (for example the possible displacement of existing activities or introduction of unfair competition) which were rarely acknowledged by appraisers and hardly ever quantified.

The equity issue is now widely acknowledged to have been neglected by appraisers during the 1980s (Pacione, 1990). Following the House of Commons Employment Committee report (1988) which criticised the London Docklands Development Corporation for failing to take account of the impact of its programmes on local residents, some UDCs have paid more attention to the likely beneficiaries of its actions. However, community oriented projects have continued to be seen as exceptions to the general rule rather than signalling any fundamental shift of emphasis.

Roles and responsibilities

Many appraisers were uncertain about their roles and responsibilities. In some cases this led to gaps in the appraisal process and some projects bypassed rigorous appraisal as a result. Conversely, in other cases, the confusion regarding roles led to unnecessary duplication with projects being 'over-appraised'. Thus, for example, projects put forward for Urban Development Grant (the forerunner of City Grant) were often appraised three times – firstly by local authorities, secondly by staff at the DoE's Regional Offices and thirdly at DoE headquarters (Martin, 1990). This inevitably delayed the implementation of projects and contributed to a considerable underspend (Pearce, 1988).

Appraisal skills

Considerable efforts were made to ensure that formal written guidance was provided to appraisers by the 1990s. However, its quality (in terms of both comprehensiveness and clarity) varied between programmes. Furthermore, the use which was made of it also varied because of differences in the skills and experience of appraisers both within and between programmes. In the case of City Grant and the UDCs project appraisal was normally undertaken by trained staff within specialist 'appraisal teams' who were able to draw upon a range of technical skills acquired in the private sector. By contrast DLG and UP projects were generally appraised by DoE Regional Offices. Whilst staff at regional level often had detailed local knowledge they were usually relatively junior, lacked previous experience of appraisal and had received very little formal training.

Appraisal routes

In the past there has also been some uncertainty regarding the most appropriate sources of programme funding for particular projects. There is evidence that because UDG was widely seen as having more exacting appraisal procedures (particularly as regards additionality) than DLG or the UP, some well-informed private sector developers packaged their projects so that they were eligible for DLG or UP assistance and avoided UDG appraisal. This particular 'loop-hole' was plugged in 1988 by the incorporation of UDG and private sector DLG into City Grant and the formulation of clearer guidelines about the distinction between City Grant projects and those to be considered for the UP. However, similar problems were experienced in the early 1990s during the development of the City Challenge initiative which derived its funding from seven separate programme budgets (Martin, 1993).

Performance management systems

A final, technical weakness found in each of the urban policy appraisal systems has been the lack of a comprehensive system of performance management which incorporates *ex ante* (appraisal), on-going (monitoring) and *ex post* (evaluation) assessments of projects. Ideally, the same criteria would have been used for all three activities. In practice, however, they tended to be seen as almost entirely separate exercises. Appraisal criteria have often been regarded as a set of 'hurdles' over which a project must progress but which have no relevance once a project has been implemented. Furthermore, the reasons for decisions to approve projects have often been poorly documented and it has, therefore, been virtually impossible to check afterwards whether the appraisers' expectations were achieved. Combined with the rapid turnover of staff in the DoE, local authorities and UDCs, this led to a widespread lack of accountability for ensuring that appraisals were accurate and that, where they were not, lessons were learned about what particular types of projects could be realistically expected to achieve. The system of performance 'milestones' for City Challenge (Department of the Environment, 1991) represents a welcome attempt to address the discontinuity between appraisal, monitoring and evaluation. However, it is too early to judge whether this will lead to a more integrated approach to performance measurement.

Strategic Weaknesses

In addition to the technical problems identified above, a major weakness of 1980s urban policy appraisal systems was their failure to incorporate a strategic dimension. There was a tendency to focus on 'micro-issues' (for example, ensuring that materials were procured at the lowest possible unit cost), with many appraisers not even attempting to assess the likely effectiveness of projects in achieving wider urban policy objectives. Blame for this cannot be laid wholly at the feet of the appraisers, since it was the almost inevitable outcome of the way in which urban policy developed in the early and mid-1980s. As Solesbury (1987) observed, there was 'a shift from planning to opportunism as an appropriate style for tackling urban problems'. The desire to achieve short-term solutions and a reliance upon market-led principles, coupled with a concern to achieve 'value for money', influenced each of the government's major urban policy initiatives. Similarly the neglect of sustainable urban regeneration (Parkinson, 1989) and an 'experimental' approach (Thornley, 1991) were not conducive to the development of appraisal criteria which incorporated consideration of the wider context in which individual projects and programmes were operating.

A concern to exert downward pressure on public expenditure throughout the 1980s also acted as a disincentive to think strategically. Appraisers were urged, above all else, to ensure that projects received no more than the minimum level of public subsidy required in order to enable them to go ahead. As Hambleton (1988) notes, 'in such an environment, arguments over cash come to dominate the whole policy-implementation process and horizons inevitably tend to narrow'. Accordingly many appraisers simply disregarded the larger, longer-term picture.

The adoption of a strategic approach which emphasised long-term outcomes and benefits rather than short-term costs was also precluded by uncertainty about the level of funding for regeneration initiatives. Faced with a situation in which local authorities were unsure about the level of resources they would receive from central government from one year to the next, many were reluctant to devote the staff and resources necessary to develop long-term DLG and UP strategies. This also encouraged them to favour projects which would achieve spend within a one-year time horizon rather than risk backing those which were less easy to implement but which might make a greater, long-term contribution to regeneration (Martin and Pearce, 1993). Similarly, private sector developers, whilst sometimes willing to invest in one-off schemes, were often wary of participating with the public sector in developing longer-term strategies for inner city areas.

Finally, the adoption of a strategic approach to urban policy was undermined by the proliferation of separate initiatives and of the number of agencies involved in regeneration. New programmes were introduced in an incremental fashion and there have been few signs, until recently, that central government has recognised the interdependence of policy actions. Rather, it tended to focus upon single, 'one-off' issues and areas. Moreover, at the local level, the erosion of local authority budgets and the transfer of responsibilities to new agencies, including UDCs and Training and Enterprise Councils, made it increasingly difficult to achieve co-ordinated action (Martin, 1994).

Conclusions

This chapter has identified the need to improve several technical aspects of urban policy project appraisal systems. In particular it has highlighted the way in which appraisal has often been perceived as a narrow, bureaucratic procedure focused on the legitimising of decisions. Comparatively little attention has been given by appraisers to alternative actions and far too much weight has been placed on cost minimisation. Moreover, despite the fact that, in the long term, less tangible outcomes often prove to be most important in promoting regeneration, appraisers have tended to identify only the more easily quantified immediate and intermediate outputs of projects.

There can be little doubt that this approach has distorted the overall profile of government assistance to inner city areas. It has led appraisers to overlook projects with the greatest long-term potential to promote sustainable regeneration in favour of schemes that were easier to implement and thus promised quick results. It has also led to a failure to take account of the synergistic benefits of co-ordinated action and made it virtually impossible to measure cost-effectiveness since, even if a project's outcomes can be assessed accurately, the lack of analysis of its opportunity costs makes it impossible to demonstrate that it represented better 'value for money' than alternative courses of action.

The case for a strategic approach to UK urban policy has been made repeatedly over the last decade (see, for example, Breheny and Hall, 1984). Indeed, critics of government policy have constantly highlighted the resistance to strategic thinking across a number of policy fields and the reliance on a piecemeal approach to problem solving (Brindley, Rydin and Stoker, 1989; National Audit Office, 1990; Audit Commission, 1991). At root, resistance to a strategic approach arises from an antipathy on the part of central government to 'planning' in general and is a direct reflection of the UK government's underlying priorities. Thus, whilst it might be argued that action is required to tackle some of the technical weaknesses we have identified, the application of greater rigour will not, in itself, be sufficient. For real progress to be made there is a need to strengthen the wider, strategic context within which appraisal is undertaken.

The potential benefits of adopting a 'strategic' approach to inner city policy are readily apparent. It encourages local authorities and other agencies to establish clear aims, in the context of an agreed regeneration strategy, offering them the opportunity to devise criteria against which the merits of alternative projects can be assessed. This makes the selection of schemes which reflect longer-term aspirations and therefore represent 'better value for money' in its widest sense much more likely. Strategic thinking can, therefore, help to ensure that individual projects and programmes form part of a wider whole and are able to complement other initiatives aimed at economic, social and environmental improvement. The preparation of a regeneration strategy can therefore foster collaboration and encourage partnerships between the broad range of agencies involved in urban regeneration. Moreover, it can be employed as a basis for monitoring and evaluating the outcomes of urban policy initiatives, thus enabling local agencies and central government to adopt a more informed approach to future policy formulation, resource allocation and project appraisal.

There are some signs that central government is responding to calls for the adoption of a strategic approach to inner city policies. In recent years the DoE has, for example, funded five-year rolling programmes of derelict land reclamation involving reasonably long-term commitments of DLG to a series of interrelated projects (Johnson, Martin and Pearce, 1992). More generally, the creation in England of Unified Regional Offices, bringing together four central government departments, may facilitate greater co-ordination of government programmes within the context of new regional and sub-regional regeneration strategies. Furthermore, both City Challenge and the SRB initiative (under which twenty separate programmes has been brought together under a single umbrella (Government Offices for the Regions, 1994)) stress the need to take a longer-term perspective, integrate individual initiatives within the context of regeneration strategies and promote partnerships between local agencies. However, whilst they offer greater security of funding for those areas which are successful in bidding for resources they do not represent an increase in the overall level of central government funding for inner city initiatives; allocations are simply top-sliced from existing programmes and areas that secure assistance gain at the expense of those which fail to do so.

The adoption of a truly strategic approach would require a far more radical policy shift towards a concern 'with creating the conditions whereby the longer-term desires, aims and objectives of individuals, organisations and areas can be achieved' (Roberts, 1990). This, in turn, implies that the imperative 'to get things done' would need to be replaced by appraisal systems which stress the integration of projects rather than treating them in isolation and that much more attention would have to be given to the overall requirements of an area.

There would need to be a new emphasis on:

- the co-ordination of initiatives, rather than focusing on single issues;
- a long-term commitment to regeneration rather than stressing short-term costs;
- a genuine partnership between agencies in the public and private sectors;
- and the encouragement of local regeneration strategies.

At the heart of this process is the need to develop projects and programmes which are consistent with agreed policies and to identify and implement projects which respond to programmatic need. The interrelationships between policies, programmes and projects would, therefore, need to be expressed clearly, and projects selected on the basis that their objectives were consistent with those defined at both policy and programme levels.

In the 1980s ministers were opposed to the adoption of a strategic approach to urban policy because of its implied commitment to intervention. It might be argued that they therefore got the appraisal systems which they deserved and, indeed, desired. There are signs that fundamental, ideological objections to a strategic approach have now abated. Unfortunately, however, the strategic approach is likely to remain unattractive to central government because of a fear of being drawn into long-term funding commitments which would cut across the now annual ritual of tough public expenditure settlements. Failure to adopt a more strategic approach will, however, mean that we run the risk of simply repeating the mistakes of the last decade with inner city projects continuing to be appraised in what is, in effect, a policy vacuum.

Note

1. Urban Development Corporations are agencies funded by central government to regenerate specific inner city areas which have fallen into dereliction, generally

because of the decline of manufacturing industry. They have wide ranging planning powers and have funded major schemes involving land reclamation, infrastructure provision and property developments.

City Grant was introduced in 1988 to provide central government assistance to the private sector to 'bridge the funding gap' associated with inner city developments which would not proceed in the absence of public subsidy.

The Urban Programme was introduced in 1978 and abolished in the early 1990s. It was designed to support economic, social and environmental projects in 57 'urban priority areas' with the DoE providing 75 per cent and local authorities 25 per cent of the funding.

Derelict Land Grant was originally available to local authorities and the private sector to promote the reclamation of derelict land. From 1988 the private sector element was incorporated into City Grant.

8

DEVELOPING QUANTITATIVE INDICATORS FOR URBAN AND REGIONAL POLICY ANALYSIS

Cecilia Wong

Introduction

The derivation of successful urban and regional regeneration policies very much depends on knowledge of the strengths and weaknesses of different areas. The economic trends of recent years have left persistent urban and regional disparities in prosperity, leading to the danger that differentials in development potential between areas are increasing rather than narrowing (Lever, 1993). Academic debate over the causes of continuing uneven development has increasingly emphasised the local dimension and the distinctive mix of relative (dis)advantages possessed by each area. It is this mix which makes an area more or less likely to benefit from each distinct type of regeneration opportunity (Fielding and Halford, 1990).

For policy-makers, then, it is clearly important to assess 'best practice' in the methods for measuring the potential and the problems of individual areas to facilitate policy targeting. Improved assessment can also contribute to the co-ordination of different agencies who are carrying out local economic development programmes (Audit Commission, 1989). At the same time, central government is closely scrutinising public expenditure and monitoring the effectiveness of individual policy activities. The European Commission too, has stressed the importance of periodic assessment of the value of European assistance (CEC, 1991).

Due to the urgent need for an improved and reliable flow of intelligence to the decision-making process, there has recently been a surge of interest among policy-makers in using statistical indicators in a number of ways: (1) to measure the needs or opportunities of each area as a basis for resource allocation; (2) to set up the contextual 'baseline' of an area's conditions to help measure the additional improvement brought by public policy intervention and assistance; and (3) to help distinguish just which opportunity or problem is most important for each area.

The use of socio-economic indicators to inform policy decisions dates back to at least the mid-1960s in the United States and Britain. Unfortunately, the initially rapid development of the 'social indicators movement' suffered a set-back in the late 1970s, due to the failure of research to resolve conceptual and methodological difficulties (Carley, 1981). In order to ameliorate the danger of feeding in a haphazard collection of statistics in a 'garbage in, garbage out' approach, it is important to derive indicators in a systematic manner rather than on an arbitrary basis. This chapter illustrates the challenges faced in developing indicators, taking an example from an earlier study on urban regeneration (Coombes, Raybould and Wong, 1992). A four-step procedure, working from general to specific, is proposed here as the basis for a consistent development process to improve the quality of indicators: (1) conceptual consolidation, (2) analytical structuring, (3) identification of indicators, and (4) creation of an index.

Conceptual Consolidation

The first, and probably the most important, step in the process of developing indicators is to clarify the basic concept which is to be represented by the analysis. Many of the key terms in policy discourses (such as 'regeneration' or 'deprivation') are subject to numerous interpretations or competing theorisations, hence it is essential to clarify the content of any such concept which is to be the subject of the analysis. This is especially important if the eventual index is to be widely accepted as relevant to policy-making. Duvall and Shamir (1981) also argue that the recognition of the basic conception is very important as it will lead to different indicator systems which represent different interests.

It is a fundamental task to address basic questions, such as 'what is the purpose of the study?', 'what issues are linked to specific programme objectives?', 'what policy instruments will be used?' and 'what is the appropriate unit of analysis?', from the very beginning of the study. These questions are vital in clarifying the causal factors and the issues which decision-makers consider to be most relevant. The questions also help to specify the most appropriate spatial units for policy targeting, which will then provide the underpinning of all statistical work in the later stages. An overall review of the best practice in related research and in the policy arena should lead to a detailed discussion with the agencies involved. This process inevitably involves value judgements of policy-makers who are the end-users of the indicators or the combined index.

Turning to the specific example of regeneration which will be followed through in this chapter, it is important to clarify the issues to be analysed. The concept of regeneration has long been recognised as involving a suite of different issues; the Department of the Environment in the UK was well aware of the complexity and interconnections of these issues. This awareness explains the Department's interest in examining factors beyond the physical and environmental development issues (which are within its core policy remit) to embrace the wider context of local economic development and social regeneration. Based on this broad notion of regeneration, Coombes, Raybould and Wong (1992) identify six types of 'resources' (locational, infrastructural, human, intangible, amenity and financial) which contribute to the distinct potential of an area to benefit from certain regeneration opportunities. These six categories are put forward as a more rounded approach to assessing an area's strengths and weaknesses. Of course, this broad approach provides policy-makers with the flexibility to focus on those key issues that are directly related to a particular programme and thus facilitate future policy targeting.

European Community policy-makers have also not been slow to appreciate this diversity of influential factors for areal development. Despite the daunting problems of statistical incompatibility and incompleteness of data sources in different countries, the Commission for the European Communities (CEC, 1992) has compiled several indicators to create a 'synthetic index' for its *State of the Regions* report. More importantly, the CEC has funded research on the key factors operating to fuel the uneven development which remains endemic within and between member countries.

Analytical Structuring

After clarifying the key concepts to be measured, the second step in the index development process is to provide an analytical framework within which the indicators will later be collated (in step 3) with a clear conceptual and policy rationale. Gurr (1981) shares the same view, as he agrees that sound systems of indicators must be constructed on coherent conceptual frameworks and then purports to offer a comprehensive schema. Ward (1981) also argues that the usefulness of a social index depends on a relevant, policy focused, definitional disaggregation of social categories. It is, therefore, important to avoid any attempt to produce a multivariate index by simply combining an ad hoc collection of indicators without any systematic thought of the issues to be measured. Although a comprehensive account of the phenomenon may not necessarily be achievable in the later stage (e.g. because of data availability problems), the adoption of an analytical framework can ensure our knowledge of errors and omissions, rather than illusions of their absence (Duvall and Shamir, 1981). A well-constructed framework is vital not only to underpin the collection of statistics in step 3, but also to guide the final analysis on index creation in step 4.

There are a number of approaches to identify the relevant factors. The 'bottom-up' approach is to list the factors which can be argued to be important individually. In contrast, a 'top-down' approach starts from an a priori analysis of the concept concerned, from which a typology of factors is derived to provide a framework of the study. There are obvious advantages in combining both approaches because the framework from the top-down approach can be set against the bottom-up list of factors so as to identify any important gaps. An early attempt to recognise the range of relevant factors in relation to urban and regional regeneration was the analysis by CURDS (1979) on how areas 'mobilise indigenous potential'.

With respect to regeneration studies, a more specific categorisation of the key issues can be derived from a bottom-up approach by drawing upon a wide range of literature and policy analysis in Britain and other European countries (e.g. Cheshire, 1987; IFO, 1990) and the well-developed debate in the United States on the evaluation of *state business climate* (e.g. Boyle, 1989). Hence twenty-nine factors which may shape an area's regeneration potential, positively or negatively, are identified (see Table 8.1). This check-list of key issues is then set against the top-down framework of the six resource categories. The twenty-nine factors fall neatly into six resource types to provide a plausible framework of policy-related issues on which areas will be measured

Table 8.1   Resource categorisation of regeneration factors

Resource category	Economic regeneration	Factors associated with: Any regeneration	Physical regeneration
Locational	Telecomms	Locational accessibility	Regional context
Financial	Local linkages Local control	Cost of investment Investment finance	Return of investment Consumer demand
Infrastructural	Industrial structure R&D activity	Communications Space constraints	Housing market
Amenity	Educational facilities Welfare facilities	Health facilities Pollution & hazards	Environmental features Leisure facilities Climate
Intangible	Industrial relations	Institutional capacity Community cohesion	Place image Cost/quality of life
Human	Enterprise	Labour force	Demographics

Source : Coombes, Raybould and Wong, 1992.

by statistical indicators in the next stage. It is important to consult policy-makers in this step to discuss the significance of the factors in relation to their main policy concerns.

Identification of Indicators

The third step of index development involves the translation of the key factors into specific measurable indicators. The policy issues identified earlier in step 2 now provide a framework from which a wide range of possible indicators can be sought. The starting point of drawing a 'wish list' of indicators is usually from an extensive review of related policy practice and academic literature. In most cases, numerous potential indicators can be identified for each key issue. This is less true once the data availability problems have been allowed to eliminate some of the candidate indicators. A single *perfect* indicator cannot usually be found to adequately represent each issue; the available data is more often in the form of proxy measures. This leads to a strategy based on a broader set of indicators, so that the analysis can draw upon a more broadly based set of measures. Recognising the imperfection of the data available means that the selection of indicator has to be rigorously assessed. Structured assessment of the value and practicalities of each potential indicator has to address five basic criteria: data availability, geographical specification, time-series prospects, implementability and interpretability.

Data availability

Data availability is one of the most fundamental problems restricting the eventual set of indicators. A considerable proportion of the relevant data for urban and regional development is not in the public sector. Many of the potential key government sources (e.g. the Labour Force Survey) are from samples whose size is too small for local analysis. Coombes, Raybould and Wong (1992) found that over half the indicators proposed to assess urban regeneration potential require non-government data sources.

Geographical specification

Though some data series are available, they may be of incomplete *coverage* of the spatial area of concern. For example, the Tyne and Wear Joint Information System provides an ideal land use database for the operation of regeneration indicators (Spicer and Grigg, 1980). Unfortunately, as only Tyne and Wear is covered, this database can only serve as a pilot for innovative measures. The spatial *resolution* of the available data may not be at a sufficiently detailed spatial scale to give adequate intelligence for some analyses. On the other hand, some input data is available for smaller areas than the output spatial unit which has the best analytical value. For instance, Coombes and Raybould (1989) chose the wider context of local labour market areas as the optimum spatial unit in their analysis of small firm formation despite the availability of data at the sub-district level.

Time series prospects

The information sources identified for the indicators will ideally be updated on a regular basis, so that they can be monitored through time and provide a dynamic analysis. However, different information sources are updated with varying frequency, and this often affects indicators which draw upon data from more than one source. One general problem of developing indicators is the tendency of over-reliance on Census information which is only updated every ten years. This problem is more obvious when small area statistics are required. Although the 1991 Census will provide tremendous opportunities for indicator development, the cycle of data obsolescence will return in the long run. The way forward is to diversify the sources of data used to give a more contemporaneous picture.

Implementability

The implementation of some indicators is relatively straightforward, though it may be tedious (e.g. requiring some kind of spreadsheet calculation). In some other cases, the indicator only becomes available after lengthy primary data collection or complex compilation and processing of the original data. Some indicators also require Geographical Information System (GIS) inputs or software that is not in the public domain.

Interpretability

The question of interpretability is the single most important part in the evaluation process, because the objective of developing indicators is to provide measures which adequately reflect the key issues of concern. Due to the difficulties of obtaining direct measures in many cases, there will be the need to develop proxy measures which demands more rigorous validity checks. Other problems of measurement reliability can also affect the interpretability of indicators. Inspection of a potential indicator's statistical properties, for example, might show that it varies so wildly between neighbourhoods as to cast doubt on its reliability.

The implementation of the above five appraisal criteria should be carried out within a structured schema. Table 8.2 gives an example of an evaluation framework used to assess the indicators of urban regeneration potential (Coombes, Raybould and Wong, 1992). The example illustrates how the three indicators for the *cost/quality of life* factor, proposed in the study under the heading of *intangible resources*, were evaluated. The first indicator, *cost barriers to first-time buyers*, performs well on all criteria though it requires mortgage transactions data from building societies. The *insurance costs and crime risk* indicator is a newly developed indicator which appears to be robust and meets all other required criteria satisfactorily. Nevertheless, its application requires further sensitivity testing to determine its value. The final indicator on *quality of life valuation* benefits from the completed analysis of the Glasgow 'Quality of Life' study (Rogerson *et al.*, 1989). However, the spatial unit of analysis is based on 72 local labour market areas (LLMAs) which do not cover certain smaller Urban Programme districts, while some districts are not distinguished where they form part of larger LLMAs.

Creation of an Index

The final stage in the development of an index is to synthesize the proposed indicators, according to their relative importance, into a single measure which will be used for policy targeting. One possible approach is to focus on the key factors derived from the analytical framework in step 2. If a single most representative indicator can be identified for each key factor in the framework, the issue of weighting can simply concentrate on the relative importance of each factor without the need to consider individual indicators. However, practical problems such as data availability as discussed above usually impose constraints on the selection of indicators and their quality. It is, therefore, rarely possible to find indicators that can perfectly represent the key factors of the analytical framework. Because of this limitation, it is important to examine the properties and the reliability of individual indicators in the process of creating a combined index. The initial step is to undertake some statistical exploration of the compiled database. For instance, the very similar statistical patterns exhibited by two indicators (which supposingly represent two different issues) may possibly imply one or both of these indicators are poor measures of the key factors concerned. Data validation is thus considered to be a prerequisite before we turn to the challenge of seeking a weighting method to create a multi-variate index.

Non-statistical weighting methods
The first category of methods is characterised by their simplicity, i.e. that the

Table 8.2 Evaluation of indicators of locational resources

Indicator	Availability		Geographical		Time series	Implementation	Interpretability	Application and opportunities
	Source	Form	Coverage	Output				
Cost barriers to first-time buyers	0	**	**	**	**	**	**	W
Insurance costs and crime risk	*	*	**	**	**	**	*	S W
Generalised 'quality of life' valuation	*	**	0	*	?	**	*	S X

Key:

	Source	Form	Coverage	Output	Time series	Implementation	Interpretability	Application and opportunities
**	all local data from GSS	all local data from on-line databases	UP areas all individually covered	available for wards /postcode sectors	robustly updatable at least annually	easy application e.g. spreadsheets	known in the literature: robust	C: consider collection of new data
*	all non-GSS data is openly marketed	easy manual data input is needed	some UP areas are combined at source	output is 'towns' – this fits the issue	robustly updatable every 2/3 years	a manual input so may take more time	appears robust – consult or test	D: local data needed from other government departments
0	some data requires to be negotiated	extensive data collation is needed	not all UP areas covered by data	areas are too large for this issue	issue is more dynamic than dataset	major GIS or CURDS software needed	the pilot results justify more work	R: research on indicator value needed
?	no source found for required datasets	unknown until data source is found	unknown until data source is found	unknown until data source is found	unknown until data source is found	unknown until data source is found	dubious – on basis of 3 areas' results	S: indicator sensitivity to be tested
								T: talks with relevant experts needed
								W: possible within GSS
								X: external input needed

Source: Coombes, Raybould and Wong, 1992.
Note: CURDS Centre for Urban and Regional Development Studies
    GIS Geographic Information System
    GSS Government Statistical Services
    UP Urban Programme

indicators are combined in a way which is easily understood. The advantage of simplicity is visibility, which means the decisions on weighting can be easily recognised and debated. However, a simple method is not necessarily a less contentious option because it may not provide the most appropriate answer to policy targeting.

(1) Null

The default method is not applying any weights to the selected measures. The 'Booming Towns' analyses of Green and Champion (1991) provide examples of a preference for applying *null* weights to the selected indicators. The apparent benefit of simplicity from this approach is also clearly a disadvantage, in that it assumes all indicators are of equal importance – regardless of the concept involved, the nature of the data available, or the objectives of any specific policy initiatives for which the ranking is needed.

(2) Expert

A fairly familiar method is to obtain the assessment and opinions of *experts* in the specific application field. For instance, the weighting scheme used by the Grant Thornton index to measure *state business climate* is based on the poll of the state manufacturers' associations (Boyle, 1989). The weighting scheme could be obtained by asking the experts' preferences directly, or using an iterative technique such as the 'Delphi method' (Sackman, 1974). The use of expert weightings has the advantage of integrating practical experience into the analysis. However, it is difficult to decide who are the experts and how to derive the precise weightings from their judgements. Of course, the results of this approach may also be open to criticism of involving personal values, vested interests and bias.

(3) Literature

As an alternative to relying on policy experts, the weighting values can be abstracted from the literature by reference to a respected study (or studies). For instance, the weighting scheme used by the *Development Report Card for the States* (Corporation for Enterprise Development, 1991) to combine different components into indices could be used as a basis for a single index of economic regeneration. However, it is unlikely that there will be a pre-existing study which covers exactly the *issues* identified by a particular shift in policy concerns. Moreover, these weightings would still need to be expressed in a set of numerical values for each of the indicators generated.

(4) Public opinion

A survey of the relative importance of the issues of concern may provide an objective measure of the public's overall views. For example, Rogerson *et al.* (1989) explored the public's assessment of the factors which make up the 'quality of life' available in any area. However, it is very unlikely that such weightings obtainable 'off the shelf' from an earlier study can be matched onto the indicators generated by any study which has been undertaken for a different purpose. On the other hand, because of the time and the expenses involved, conducting a new opinion survey may not be a practical option. The problem of unreliability of opinion polls also casts doubt on the adoption of this approach.

Statistical methods

Due to the practical difficulties involved in using the weighting systems mentioned above, an alternative way forward is to focus on a purely empirical assessment of the indicators themselves. Various statistical techniques can be used to produce a combined multi-variate index from the selected indicators.

(1) Z-scores

The method for creating z-scores starts by examining the statistical distributions of the raw data for each indicator: those which show a skewed distribution have to go through a normalisation procedure. Each variable is then transformed into a standard form. The standard scores on each indicator for each area are then either added or subtracted, depending upon the interpretation of positive values. The biggest advantage of this form of composite score is its simplicity which can be easily understood. It also allows policy targeting by ranking areas at a variety of different spatial levels. However, this method tends to oversimplify the data by ignoring the complex relationships between the issues which the indicators represent. It easily leads to the danger of 'double counting' when some indicators are highly correlated. Hence, this method is less appropriate for handling a large number of indicators.

(2) Regression analysis

Regression analysis provides a convenient summary of the importance of various indicators (independent variables) according to their strengths in explaining the variations of a single all important measure (dependent variable). For example, Coombes and Raybould (1989) used value added tax (VAT) data to model factors affecting local enterprise activities. The advantage of regression analysis is that it could be used for description of the dataset analysed as well as predicting the outcome of the wider population or at a different spatial scale. The regression coefficient of each independent variable provides an automatic weighting on the dependent variable which they seek to explain. However, the biggest problem of this method is finding a single valid variable to represent the concept in a suitably rounded way. Ideally the choice of variables used in the model should be theory driven; however, in most cases they are purely based on the past experience of the analyst. Also, there are limitations of using regression models for prediction by assuming that the current model is still valid for the predicted observations.

(3) Factor analysis

Factor analysis is used to identify a relatively small number of factors which can be used to represent relationships (by explaining as much variance as possible) among sets of many variables. An example in the policy field is the Duguid and Grant (1983) use of factor analysis to combine several indicators into a single deprivation score to prioritise areas of special need in Scotland. One of the strengths of this technique is that the obtained factor(s) help clarify the general concept on the basis of the empirical links within a set of indicators. It also provides an automatic statistical weighting of each variable on the factors, hence factor scores obtained for each areal unit can be used for ranking. Equally, factor analysis can be seen to have some disadvantages. The application of factor analysis involves critical decisions, such as which statistical options should be used in the statistical procedures, and which and how many factor(s) should be used for ranking. The process of assigning a label to each factor according to their attributes is also highly subjective.

(4) Multi-criteria analysis

The results from a multiple factor analysis cannot yield a single ranking solution on their own. However, they *can* provide the basis for a multi-criteria analysis. The factor scores for the chosen factors for each spatial unit can be assessed to see which exceed a threshold value on a set number of factors qualified. Massam (1993) describes several versions of this method to illustrate the ways in which 'spatial coincidence' of several factors can contribute to policy-related analyses and decision support. The strength of this method is that it can be closely linked to policy concerns through, for instance, distinguishing which areas score highly on which particular factors. However, the operation of this method requires lengthy and complex explanation; and no simple ranking can be calculated for the individual spatial units, although it is possible to rank an upper tier set of areas (e.g. local authority districts) on the basis of the proportion of their population which live within those lower areas (e.g. Census enumeration districts) that fall into the target categories.

(5) Cluster analysis

Cluster analysis is a statistical technique which aims to classify areas into relatively homogeneous groups. This method has been widely applied in the private sector to create area classification schemes as a means of discriminating variations in consumer behaviour (e.g. the Charlton, Openshaw and Wymer (1985) analysis to define Super Profiles). The characteristics of each cluster can be identified from the descriptive statistics of each variable. This method can provide a very parsimonious solution by identifying the target areas in just a few clusters and it takes into account the different dimensions of the issues concerned within the classification process. Equally, there are notable disadvantages of cluster analysis as it requires detailed and debatable operational decisions throughout the whole statistical procedure. Firstly, the measurement of some form of association or similarity between the areas is needed in order to show how many different groupings really exist in the study. It is then up to individual researchers to determine how many outcome groups they would like to obtain. The next step involves the profiling of the areas in order to determine their composition and to facilitate the labelling of each cluster. Also, the identification of which clusters should be considered to be the 'target' areas for a particular policy is based on the judgement of policy-makers with respect to the characteristics exhibited by different clusters. No ranking of individual areas can be obtained as an area is either 'in' or 'out' of the chosen cluster(s).

Any recommendations as to which method is best in producing a combined index have to reflect the balance between simplicity, statistical robustness and flexibility. The obvious advantage of adopting a simple method is to avoid a statistical blackbox and make the process transparent for interpretation. With respect to the latter two criteria, some of the statistical options for obtaining weightings can be seen to be unsatisfactory on the grounds that they are either impractical or so arbitrary as to be potentially contentious. *Factor analysis* emerged as the preferred option in the recent methodological reviews by Bartholomew (1988) and Bell (1990) as the most robust approach to combining indicators. However, the application of this technique has to be cautiously handled and very much depends on the nature of the compiled database. Hence, it is important to carry out a preliminary validation analysis of the assembled database to identify the differences in the outcome produced by alternative approaches before making the final judgement.

Conclusions

This paper aims to raise a number of methodological issues involved in developing quantitative socio-economic indicators. A four-step procedure has been outlined to provide a consistent and coherent framework to guide the development of targeting and ranking analyses and to avoid the danger of creating a haphazard collection of intuitively acceptable indicators. However, the claim here is not that the many problems encountered in developing indicators have been resolved. In fact, there is rarely a simple 'right' or 'wrong' approach, but there are more 'appropriate' solutions which can be found if the best practice points for producing a multi-variate index are borne in mind. Five specific issues are raised here to guide the development of indicators:

(1) The basic nature of socio-economic indicators is that they are not value-free. It is inevitable that value judgements from policy-makers and researchers will be involved in the process of defining the conceptual issues and developing a policy relevant analytical framework. This form of judgement is obviously reflected from the explicit numerical values which can make up a form of weighting to produce a combined index. As it is impossible to be value-neutral, it is essential to include policy-makers as much as possible in each step of the development process, otherwise the exercise is likely to remain on the drawing board.

(2) As socio-economic indicators tend to be measurable surrogates for some unmeasurable or latent concepts, such as regeneration potential and enterprise culture, the ideal solution will be more likely to arise if there is a well-established theoretical understanding which can underpin the rationale of the choice of certain indicators. Hence, theoretical development is urgently needed to examine the causal relationships of different elements for urban and regional development. Collaboration between social scientists from a wide range of disciplines is required to enhance a comprehensive study of many issues in the diverse context of relevance to urban and regional development.

(3) The production of an appropriate multi-variate index is rooted in the need for an initial exploration of the database of indicators and to identify which features are crucial for any specific application.

(4) It is important to urge better practice in public data compilation (Wong, 1993) because indicator research always faces the set-back of poor data availability, updatability, spatial aggregation and patchy spatial coverage. Equally, researchers have to explore data sources which are not in the public domain; such sources will become more and more important with the commercialisation of information and the reluctance of central government to compile data for which it does not have an immediate policy need.

(5) More sophisticated analyses are needed of the data which is available. Growing use of GIS will help familiarise more users with analyses such as accessibility surfaces. The challenge for researchers is to demonstrate the relevance of their techniques and to increase the interest of policy-makers in the skills which they offer. The incentive for policy-makers is to reduce the misallocation of public money which flows from the inadequate methods that are currently used for policy targeting.

Acknowledgements

This paper arises from work undertaken on a series of NE.RRL projects at CURDS, funded mainly by the Department of the Environment and Scottish Homes. The author acknowledges the inputs of other researchers, in particular Mike Coombes and Simon Raybould of CURDS. A longer article on this topic has been published with Mike Coombes in *Environment and Planning A* (1994) Vol. 26, no. 8, pp. 1297–1316.

9

CHANGES IN LOCAL GOVERNANCE AND THEIR IMPLICATIONS FOR URBAN POLICY EVALUATION

Rob Imrie and Huw Thomas

Introduction

The last ten years have witnessed a significant change in the institutional fabric of local government and of the institutional mechanisms underpinning the development and implementation of public policies (Cochrane, 1991; 1993). Such changes are well documented and range from the selected decentralisation of service provision and funding from public to private organisations, to the emergence of a business involvement in public policy through, for example, Urban Development Corporations (UDCs) and Training and Enterprise Councils (Hoggett, 1991; Imrie and Thomas, 1993; Stoker, 1988; Williams, 1994). Such transformations have led Keating (1991), amongst others, to talk of a new closure in the politics of local governance, with the emergent privatised local state characterised by limited channels of local (democratic) accountability, a situation compounded by the fact that there are few legal requirements on the new institutions to open up to wider participatory systems (see also Eisenschitz and Gough, 1993). Thus, for example, the UDCs are under no obligation to disclose the minutes of their board meetings nor do they have to admit members of the public in the way required of local authorities.

Such institutions are beginning to dominate British public life. Yet they are characterised by inadequate scrutiny and minimal accountability while controlling huge segments of public finance. In particular, the restructuring of the welfare state, towards the utilisation of quasi-markets, is placing some emphasis on 'value-for-money' criteria in determining spending priorities leading to what Barnekov *et al.* (1989) refer to as a 'balance sheet' mentality in the new institutions. Some see this as indicative of a new agenda in the public sector, the adoption of a series of values mimicking private sector corporations and using the language and practices of commercial confidentiality to minimise external scrutiny of their operations and behaviour (Stewart, 1988). As Cochrane (1993) notes, such behaviour seems related to an increasing institutional fragmentation, characterised by the growth in service providers competing with each other, while state expenditure is increasingly being directed through private sector contractors operating with values and attitudes not necessarily conducive to critical evaluative studies of their performance and of the policies that they are charged with developing and implementing.

This chapter considers the changing institutional context within which British urban policy evaluation is taking place in the 1990s, and suggests that there are reasons for expecting increasing suspicion on the part of agencies delivering urban policy of the subversive potential of independent evaluation. The changing contextual circumstances, to which the paper draws attention, have implications not only for the conduct of research but also for the place of networking between policy-makers and policy evaluators. We begin by briefly reviewing the organisational realities of policy evaluation, then we outline, in turn, three trends which, we suggest, have altered the context within which urban policy evaluation is being undertaken: the growing influence of privatism and market values in urban policy (Barnekov et al., 1989); institutional restructuring within the state, locally and centrally (see, for example, Hoggett, 1991); and the increase in urban entrepreneurialism (Harvey, 1989). We suggest that all of these have generated a significantly different, problematical, context for the conduct of urban policy evaluation.

We develop this argument by considering the conduct of evaluative research in relation to the British UDCs. In particular, by using a case study of the Cardiff Bay Development Corporation (CBDC) we critically discuss the possible ways in which independent evaluative research is being transformed by virtue of the changing socio-institutional context of urban policy, referred to above. In doing so, we refer to three interrelated aspects of the research process, that is, establishing legitimacy within the new institutions, working within the organisational frameworks, and (re-)presenting the findings of policy evaluation. As our discussion will indicate, there is evidence of sensitivities on the part of the new institutions of local governance which are creating difficulties not only for the conduct of policy evaluation but also for mainstream academic research. In particular, the specific closures within the new institutions have, as we will show, the potential to inhibit not only the empirical documentation of the transformations in local governance, but also of the conceptualisation of the processes underpinning such changes.

Evaluation, Organisational Interests, and the Changing Context of Urban Policy

The proliferation of new policy institutions since the early 1980s has not been matched by the systematic monitoring and evaluation of their performances and policies. Indeed, Hart (1991), in the US context, notes the overall diminution in evaluative research throughout the 1980s, to the point where a crisis in evaluation had emerged with the majority of studies being inappropriate because of their short-term duration, their specific nature excluding key relationships, and the absence of external evaluation with the majority of studies being conducted in-house. One example of this, in the USA, is the marked trend towards internal (i.e. unpublicised) and informal evaluations of non-defence executive programmes (see Barnekov et al., 1989). Yet, in the UK context, Bach

(1992) has noted that, throughout the 1980s, there was a considerable amount of evaluation of Urban Programme initiatives, while others have shown that there was a progression in the Department of the Environment's (DoE) evaluation of the UDCs, from assessing intermediate output measures to increasing concern about the broader social and economic impacts of strategy.

Yet, while this suggests that the role of evaluation has maintained a position of some importance in formal, government, policy programmes, a range of authors suggest that the scope and form of evaluative research is often illogical, weak and increasingly concerned with a narrow range of evaluative criteria (Turok, 1989; 1991). Thus, the measures utilised by the UDCs have been criticised as vague, imprecise and contradictory, for not distinguishing, for instance, between the tenure of housing, or the types of jobs created, and providing physical performance targets where social and economic impacts result (Colenutt and Tangsley, 1990). Such observations reinforce Turok's (1989) disquiet with methods and approaches which primarily emphasise the quantitative measurement of policy impacts where there is little concern with imputing causality or examining how changes are produced or why they are not, what some have termed a technocratic approach. Thus, Hart, summarising Turok (1989), concludes that policy evaluation needs to 'look more at mechanisms linking policies and outcomes and the conditions that facilitate or obstruct their effectiveness'.

This also implies that the traditional focus of evaluative research may need to be reconsidered. As Turok (1989) notes, typically policy evaluation has sought to measure impacts of initiatives through large-scale surveys of recipients of assistance. Yet, if there is a need to understand better the conditions and mechanisms facilitating outcomes, it seems clear that more attention should also be paid to evaluating the specific roles and performances of the socio-institutional, or organisational, structures largely responsible for the policy initiatives in the first place (and, in some circumstances, evaluating them) without recourse to crude, quantitative, measures relating solely to a narrow range of output measures. Yet, as some documentation shows, a real difficulty facing independent researchers appears to be organisational resistance to external evaluation, especially evaluation which falls outside of the technocratic model (Bryman, 1988). This chapter will suggest that such resistance has intensified in the context of the emergence of privatised policy fora, although there is much evidence which indicates the general reluctance of many organisations, public or private, to open themselves up to external or independent scrutiny (Beynon, 1988; Bryman, 1988). This has obvious implications for evaluative research.

A range of authors note how research is problematical where it impinges on political alignments, or on the vested interests of powerful institutions, making evaluative, or any, research more or less impossible to conduct. As Beynon (1988) notes, part of the problem is that, in institutional or bureaucratic contexts, the researcher is 'a relatively uncontrollable element in an otherwise highly controlled system'. For example, Horton and Smith (1988) report on the difficulties of outsiders in evaluating the performance of a police force, a classic example of a tightly knit organisation with a pervasive ideology which emphasises its distinctiveness and the general hostility and ignorance of the rest of society to it (Young, 1991). Though Horton and Smith suggest that a strategy of increasing the 'ownership' of those being evaluated over the process can increase co-operation, they also stress the need for using independent researchers and for the publication of results if evaluation is to improve practice. This reflects a perception that evaluation, however carefully conducted, can tread on toes, or, as Platt (1987, p. 242) puts it, good social science is 'potentially subversive of any status quo'.

Others consider the contrasting expectations of research outcomes held by academic researchers and (especially industrial) organisations. The former are usually more concerned with, in Lee's (1993) terms, 'a desire to make social life translucent', or with telling a story of how it was, with the latter more concerned with prescriptive and applied findings (i.e. immediately useful to the ongoing and future operations of the organisation). Indeed, many organisations have little desire to be told anything about themselves which is not useful to their (often) immediate applied and strategic behaviour, an attitude summarised by Beynon (1988) who, in commenting on the Volvo Motor Company, cites an executive director:

> Volvo has received many requests from social scientists and research people to make studies of Kalmar. However, their research was not very helpful when we were designing the plant, so we rejected most of these outside requests . . . we don't want to find out what we did yesterday. We want to know what to do tomorrow.
> (Gyllenhamer, quoted in Beynon, 1988, pp. 22–3)

While academic research need not (and should not) preclude some engagement with prescriptive questions, part of the difficulty in establishing legitimacy in organisations is coming to terms with the socio-political culture of the particular organisation that is being investigated (see the next section). Internal political factions were certainly evident in CBDC, where we have recently been conducting research, and our evaluation somehow had to avoid being co-opted by one group over another, if only to avoid marginalising ourselves with sections of the organisation. Yet, the highly politicised nature of the organisation generated such constraints (see the next section) that we felt it appropriate at times to utilise covert observation where we knew access would otherwise be denied (because we were perceived as a threat).

Indeed, policy and programme evaluation can meet stiff resistance from interests which it threatens (Carley, 1980). We are referring here to threats to interests which arise from the structure of social or institutional relations in which a group or an organisation finds itself, rather than misunderstandings arising from a clash of cultures or perspectives that can be ameliorated by educational work, collaboration and communication between evaluator and evaluatee involving, perhaps, some adjustment of research methodologies in the light of concerns and criticisms (Barnekov et al., 1990, p. 8). Thus, organisations like the UDCs represent selective, partisan, interests and, whatever their claims, they are generally perceived as pursuing interests which reflect the values of successive Conservative administrations, pro-business, pro-market, with little commitment to the pursuit of wider social welfare strategies of the type at least pursued by many local authorities. This, then, tends to generate a defensive posture on the part of the new institutions towards evaluative research which may reject, or only accept in part, the broad parameters with which they operate.

This tendency, when taken with the degradation of the monitoring and evaluation of regeneration programmes into exercises in boosterism, is persuasive evidence of the growth of a deep-rooted resistance to external evaluation and monitoring which requires both explanations and prescription for action by policy researchers and policy-makers (see Barnekov *et al.*, 1990, p. 21). Indeed, it is clear that the activity of research is highly conditioned by the wider socio-economic and political environment, where the interplay of political ideologies, values and attitudes defines, and conditions, the boundaries of 'legitimate' social research. In particular, three emergent, contextual, factors are of increasing relevance in conditioning the nature of independent evaluative research; that is, the propagation of market values in public policy institutions, the development of new organisational fluidities, and the rise of urban entrepreneurialism. We briefly discuss each in turn and draw out some of their implications for the conduct of evaluative research.

Market values

The new policy institutions are adopting many of the attitudes and values associated with privately owned enterprises in a market economy. For instance, that it is both operationally necessary, and defensible, that external scrutiny is minimised, being restricted to discouraging fraudulent and dishonest activity. In addition, the primacy of the private sector is also being extolled in terms of the perception that organisations operating in the market are somehow characterised by (the positive attributes of) dynamism, hard-headedness and clarity of purpose, virtues which, so some argue, should be transferred to agencies involved in the delivery of urban policy (Clarke and Stewart, 1994). In both the USA and the UK, there has been a growing acceptance of such values amongst urban policy-makers (Barnekov *et al.*, 1989). One reason for this is, simply, that for well over a decade governmental agencies have been encouraged to work in 'partnership' with the private sector in delivering and, latterly, devising urban policy at the local level.

As Reade (1982) notes, such partnerships constrain governmental agencies to work within private sector conceptions of 'proper' modes of working, including, in the cases he described, notions of commercial confidentiality. Indeed, it seems that the entrenchment of the virtues of competition within the delivery of urban policy will consolidate the trend towards minimising the divulging of information of potential usefulness to 'competitors' (who may be other government agencies), with inevitable implications for evaluation of urban policy initiatives (Parkinson, 1993). More subtle, and perhaps more difficult to document, is the effect of a consistent refrain in governmental rhetoric in the 1980s which has extolled the virtues of action, of *doing*. As Thornley (1991, p. 46) notes, Michael Heseltine's accusation of jobs locked up in filing cabinets, through to the establishment of enterprise zones, garden festivals and the UDCs, have all served to discourage those involved in policy from giving any priority to rigorous evaluation.

For instance, when, some years ago, one of us presented a paper on the evaluation of CBDC to a group of planners, one of the first comments in the ensuing discussion was that UDCs were action-oriented and could not devote time to collecting data (Thomas, 1989). Thus, we see here the manner in which the pervasiveness of a 'can-do' philosophy can not only influence attitudes towards evaluation, but also provide a justification for, or legitimation of, resistance to requests for co-operation in conducting evaluation.

Organisational fluidity

Urban policy is being delivered in an environment of increasing organisational fluidity where relationships between and within organisations, governmental and non-governmental, are regularly being renegotiated and redefined (see, for example, Hoggett, 1991; Stoker, 1990). The findings of evaluative studies become a part of this process, and the studies themselves are likely to be promoted (or resisted) by those with an interest in organisational change. Indeed, it seems clear that resistance and suspicion is likely to be in inverse proportion to the influence of those being evaluated over the terms of reference of their evaluation. For instance, the study of police work by Horton and Smith (1988), mentioned earlier, exemplifies this, as do the debates in various policy arenas over the compilation of statistics for 'league tables'. Thus, researchers funded by organisations which have no 'purchase' over, or links with, the agencies being evaluated are likely to arouse considerable suspicion which they have no formal mechanisms for reducing (though, of course, a part of the researcher's skill will be to defuse situations which are potentially damaging for the course of the research; see, for example, Beynon, 1988).

In addition, many policy organisations, like the UDCs, are increasingly operating with fixed-term contracts often oriented around single-task projects, while the contracting-out of policy functions, especially to private sector organisations, is increasing the range of participants in the policy process. The emergence of partnerships is also part of the development of new organisational contexts for policy development and implementation, so compounding the institutional complexity facing the evaluative researcher. As Bresnan (1988) notes, the difficulty in studying what he terms the 'variable organisation' relates simply to the situation that the type of setting is not only temporary but also subject to 'considerable change as circumstances develop through the various stages of the project cycle' (p. 48). This raises a series of problems for evaluative research. For instance, how should the researcher deal with the variable boundaries of the organisation, the propensity for its internal structures, its attitudes, policies and values, to change rapidly in the course of the development of particular projects and strategies? There is, therefore, some need to capture the dynamics of rapid change which seems beyond most technocratic, if not all, forms of policy evaluation.

Urban entrepreneurialism

A significant part of the changing context of urban policy is the local state's shift from broad-based social and welfare policies towards the pursuit of strategies seeking to promote economic growth (Cochrane, 1993). In particular, the truncation of local state finances, coupled with a diminution of central state support for a range of local welfare programmes, has heightened the local state's dependence on capturing private capital as a complementary source of finance for its projects. Harvey (1989) has described the turnaround in the practices and policies of local states in the 'West' as analogous to the business entrepreneurs, self-seeking strategists, aiming to sell the city as one would a product. Thus, entrepreneurialism involves imagery or 'the selling of the city as a location for activity which depends heavily upon the creation of an attractive urban imagery' (Harvey, 1989, p. 13). Wilkinson (1992) documents the astonishing variety of logos, advertisements and images developed by agencies 'selling' Newcastle-upon-Tyne, targeting a variety of audiences differentiated spatially and functionally. The creation of images is, therefore, a complex affair into which evaluative studies can intrude in multiple ways.

Barnekov et al. (1990) in the USA and Coulson (1988; 1990) in the UK identify perhaps the most common way in which urban entrepreneurialism impinges upon evaluation – an image of increasing prosperity, of being an area 'on the move', is bolstered by unqualified statistics about 'job creation' which is the sum total of the 'evaluation' required by policy-makers and implementors. Loftman and Nevin (1994) and Robinson et al. (1993) provide case studies which illustrate the inadequacy of the data quoted by promoters of recent urban renewal projects for a rigorous evaluation of their methods, and the former note, too, the vigour with which agencies defend their activities if their value is questioned. The need for 'good copy' also underlies the way in which studies such as Findlay, Morris and Rogerson (1988) on quality of life indicators are seized upon by publicists of cities ranked highly, and exploited uncritically. The examples are legion. More generally, competition is associated with an aversion to bad publicity and a premium being put on loyalty.

External evaluation which detracts from the positive images and messages is, therefore, to be discouraged as something which is somehow working against the good of the city as a whole. Indeed, some studies of higher education and health indicate that an increasing emphasis on competition has led to severe discouragement of 'whistle blowing' (Meikle, 1993; Toynbee, 1993). To the extent that independent evaluation, particularly of the important qualitative variety, may depend upon frank appraisals by 'insiders', this tendency has worrying implications for its feasibility (Coulson, 1990). Given this, we now turn to a fuller discussion of the pragmatics, possibilities and problems associated with evaluating the policies of the British UDCs.

Evaluating the British Urban Development Corporations: an Example

In this section we present a study of how the trends outlined above have influenced our relations with one of the British UDCs, CBDC, and, in particular, the conduct of a two-year Economic and Social Research Council (ESRC) funded project evaluating aspects of their activities. It begins by locating CBDC within its local political and institutional context, relating this to the broader considerations already discussed. We then consider three aspects of researching CBDC which illustrate some of the difficulties in investigating the new systems of local governance. Firstly, establishing legitimacy (or one's research credentials) and securing access to the organisation; secondly, working with the organisation and within its internal and external networks;

finally, exiting the organisation and disseminating the findings of evaluative research.

Established in April 1987, CBDC was promoted as a fresh and dynamic agency, but it joined an array of others which had experience of urban renewal and economic development in Cardiff. For instance, in the 1970s and 1980s, Cardiff City Council (CCC) had co-ordinated a highly esteemed city centre redevelopment; in the late 1970s, and early 1980s, the Welsh Development Agency (WDA) had led the task of redeveloping the 165-acre East Moors Steelworks site in south Cardiff, and, most recently, South Glamorgan County Council (SGCC), and the Land Authority for Wales (LAW), had, with the developer Tarmac, taken on the task of redeveloping semi-derelict land in the docks for a mixture of uses (Imrie and Thomas, 1993). Though SGCC, in 1986, invited the Secretary of State to declare a UDC in South Cardiff, its own high profile in urban renewal and economic development before and since, together with the evidence of interviews with officers and councillors, suggests that it was accepting (and seeking to influence) the inevitable. Moreover, while the city council has never officially accepted the rationale for establishing a UDC, it has consistently sought to co-operate with it at officer and political levels.

Underlying overt declarations of co-operation and consensus, therefore, are institutional tensions and rivalries, complicated on the part of the local authorities by imminent local government reorganisation which will see one of them disappear (Thomas, 1992). From time to time, the tensions became visible. Thus, for many years, SGCC and CBDC ran separate campaigns to woo inward investors, while the city council has felt it necessary to remind CBDC that it is the legitimate (elected) representative of the population of the docks area (Cardiff City Council, 1988, para. 8.5.7.1, p. 57). Meanwhile, in the wider sub-region of south-east Wales, there has long been political concern that CBDC is receiving all the available public investment. Nevertheless, it does exist, does have a large budget (1994: £47m) and continues to enjoy political support from central government. As a result, it has considerable local influence, and organisations such as the local enterprise agency (CAVE), local authorities and local business organisations have no desire to fall out with it (Imrie, Marshall and Thomas, 1996).

CBDC's strategy, to transform 2,700 acres of docklands, depends upon attracting massive amounts of inward investment (see Thomas and Imrie, 1993). It is firmly in the business of urban entrepreneurialism, creating a new image for its area which will attract not only major investors, but also local and regional visitors, and create a basis of city-wide support for its vision of the future (Thomas and Imrie, 1993). Thus, central features of CBDC's view of the world are the perceptions of south Cardiff by a range of audiences (local, but especially national and international, political as well as economic), and also its recognition that there are a number of organisations which, while not wishing to see any ill befall the city, would not be averse to seeing CBDC fail. It seeks to sell Cardiff Bay, and to sell itself, therefore, in a somewhat hostile world, and a hostility, especially at the local level, that has failed to be convinced by many of the UDC's policies, especially its plans for the barrage across Cardiff Bay and the relocation of indigenous businesses (Imrie and Thomas, 1992).

In such circumstances, we 'entered' Cardiff Bay to conduct a two-year longitudinal study aimed at evaluating a number of aspects of the UDC and its operations. Foremost was how CBDC's plans and strategies were affecting the material operating conditions of local, indigenous, firms already established and operating in the redevelopment area, yet threatened by CBDC's plans to relocate them. We were especially interested in the political interplay between the UDC and the small businesses, and how the UDC was seeking to reconcile its push for attracting global corporate capital into Cardiff Bay while maintaining, if at all, some commitment to the existing bedrock of local firms, a commitment for which the local authorities were pushing. In addition, the research sought to evaluate the firms' responses, coping strategies, and political actions towards the UDC, to evaluate the articulation of the local 'business voice' in influencing the content of urban renewal policies (see Imrie, Thomas, and Marshall, 1995).

We also wanted to know something about outside institutional pressures on CBDC, how other agencies and influences were able (if at all) to transform its policies. In this sense, we were interested in questions related to the emergence of policy and political networks and their influences on the evolving strategies for the renewal of Cardiff Bay. This remit required a different approach to policy evaluation than that offered by the technocratic model and it was clear that detailed access to the UDC was a prerequisite. We now devote the rest of our discussion to the specificities of gaining access to CBDC and attempting to work with (in) the organisation to facilitate the end objectives of our research project.

Legitimacy, consent and evaluation

For CBDC, our specific remit was problematical given that its decision to relocate up to 300 local businesses had been made, the instruments and approaches decided on, and any involvement on our part seemed to provide few practical inputs or benefits to the process as far as the development corporation was concerned. In particular, one part of the difficulties we faced, in establishing some legitimacy with the UDC, was our concern with some of the social and distributive costs and consequences of the policies that CBDC was pursuing. This was a lead issue at the end of the 1980s, with the local press reporting anger by local firms, while the city council was trying to place the issue high on the political agenda. All of this was being compounded by pressure groups campaigning to stop another part of the strategy, the building of a barrage across Cardiff Bay. So, given the politicised and sensitive nature of the issues, and what seemed to be a beleagured development corporation, it is likely that we were seen, at the outset, as posing a threat to the UDC. For us, in a context of apparently divergent interests concerning the nature of policy evaluation, how was legitimacy established by the research team, if at all?

Our first questioning came well before the start of the ESRC project, in 1989, at a meeting we had arranged with an officer of the development corporation to discuss potential research or consultancy we might undertake on behalf of it. We were somewhat taken aback to find ourselves being examined quite aggressively on whether we had ever written about south Cardiff, whether we were critical of the agencies involved in renewal there, and whether we were (as senior personnel in CBDC apparently believed) 'communists'. The clear implication was that if we were not *for* them, then we were *against* them, and criticism we had made of the LAW in a previous publication apparently persuaded some that we were not for them (the LAW and CBDC share a chairman) (Thomas and Imrie, 1989). At this juncture, we realised that the problems of negotiating access really required some redressing of our image in the UDC's eyes; not an easy task.

The pattern of hostile questioning in negotiating access had permeated a range of institutions by the time we started the ESRC research, and it seemed to us that a range of 'cosy partners' had emerged, with CBDC dominating, or, as we sometimes felt, intimidating, some of the key players involved in the urban renewal programmes. Thus, in 1992, when we began work on our ESRC funded project, we sought a meeting with the chairman of the newly formed Cardiff Bay Business Forum (CBBF). This organisation was (and is) seeking to secure a share of the benefits of urban regeneration for existing Cardiff Bay firms. As part of its strategy it has been anxious to establish itself as a constructive critic of CBDC, fundamentally in agreement with its urban regeneration strategy. In early June we arrived for our meeting to be confronted by a trio of CBBF's officers. It emerged that one of their fellow executive officers believed that we were not approved of by CBDC, and, if this were so, they would have nothing to do with us. Though a phone call to a contact in CBDC had failed to damn us, they still felt it best to question us in some detail about our backgrounds, previous publications about south Cardiff, and the current project before even answering factual questions about the history and origins of CBBF.

We have had similar episodes with CAVE, where an interviewee was convinced that one of us (who, it so happened, was not present at the meeting) was not liked by CBDC, and so refused to answer any but the most banal questions, and subsequently refused follow-up interviews with us. Though we have retained contact with the organisation it is clear that some officers remain deeply suspicious of the purpose to which our work will be put, and are, in effect, refusing to co-operate with us. It is plausible to suggest that these experiences reflect the particular patterning of power relations which have evolved in the city since 1987, with the UDC dominating a range of policy institutions, or at least tying them into particular ways of acting and behaving. CAVE, for example, is employed by CBDC to liaise with small firms in the area and is, in effect, an agent of the UDC, although, technically, it is an independent company. Yet, CBDC has given CAVE a lucrative rolling contract to manage the relocation process which, as CAVE has acknowledged to us, is vital to its material well being. It is hardly surprising, therefore, that specific limits were placed on the divulging of information to us, and only that which was seen not to compromise CAVE's relationship with the UDC.

Throughout all of this we adopted a stance which Bresnan (1988) refers to as methodological pragmatism where time and again we stressed our neutrality and independence. Our receipt of the ESRC grant, however, was rarely acknowledged as a 'badge of independence', and, over the course of the research, our legitimacy was always at 'knife-edge', a case of treading carefully. We tried a number of strategies to attain the confidence of the UDC, including dialogue, a representation of previous research, laying, as it were, our academic credentials on the table. We stressed the positive aspects of CBDC knowing what its customers (i.e. the relocatees) were thinking, and how we would be able to provide information about the impact of its relocation policies.

While much of this was like a 'hard sell' approach, the UDC, in the main, was unmoved and unconvinced, tending to regard our position as tangential, even irrelevant, to its main concerns. While this was probably an accurate reflection of our status, as far as CBDC was concerned, it created practical problems for us of securing access to relevant data and officers within the UDC. There was a danger, of which we were acutely aware, that poor access to the UDC's perspective would unwittingly distort our overall conclusions or, at best, force us to leave lacunae in our analysis of what was happening in the docklands. We took active measures, therefore, to improve our access to CBDC.

As part of this process, we institutionalised a particular pattern of meetings and feedback with officers in the UDC. Over the two years, regular meetings were conducted with the key officers responsible for land management, marketing, and the business relocations from our study area. After each meeting, transcripts would be sent back to the relevant officer and, after comment and annotation, sent back to us. At the following meeting, details of the reported evidence would be debated. While this tends to conjure up a picture of reciprocity and mutual understanding, the process was underpinned by a clear sense of what the UDC understood, or defined, as legitimate, in distinction to problematical or false, representations of the organisation. Thus, on most occasions the UDC would come back at us with a censoring of our records of meetings suggesting that its views were being misconstrued. As one of the officers in CBDC commented after the first meeting, 'the write-up is 70 per cent against us, 30 per cent for us', somehow denoting that this was problematical and not how it should be, whatever the particular 'facts' of the specific situation under investigation. It was clear, therefore, from our meetings, that the price of securing legitimacy was no negative presentation of the UDC's operations, a price that could not be paid.

Working with (in) the organisation

Any form of evaluative research is critically dependent on access to information. Yet a key characteristic of the UDCs, and other quangoes, is their resistance to external scrutiny, and their withholding of information to external bodies (Coulson, 1988; Oatley, 1989). This was a particularly problematical part of our research programme given that we were keen to establish causal relations and mechanisms, to try and understand what was leading to the specific effects of regeneration and renewal in Cardiff Bay. In particular, our interest in questions relating to the leverage of political power, of policy networking, and of levels of policy influence, meant that our evaluation of CBDC required much more than single interviews with officers in the UDC, but, crucially, access to documents, files, and other items which would permit a detailed picture of intra and inter-institutional networks. Yet from the outset it was made clear to us that access to the organisation was to be restricted to a relatively small number of senior persons, with written and other non-verbal forms of information prohibited to us.

This, of course, posed problems in the conceptualisation of the processes under investigation, and of the organisations who were the focus of our attention. Indeed, one of the real ironies of social research is that limited access and collaboration has the potential to generate weak or inadequate, even unfair and misleading, conceptualisations of socio-political processes. Yet, faced with a context of limited information (which is always the case anyway), researchers have no option but to theorise on the basis of what they have. As a CBDC officer commented, after the presentation of some of the findings about firms' attitudes towards the UDC, 'this is really unfair and doesn't show us how we really are'. Yet, as we pointed out to the officer, his organisation had been actively involved in the 'production' of the information on which, in part, our assessment had been based, or, more accurately, its withholding of key information had reduced the information base around which we were able to construct our evaluation. Indeed, we had asked them to comment on the views and attitudes of the firms and to provide counter evidence, yet this was dismissed by the organisation. The views of the firms, as far as CBDC seemed concerned, were an irrelevance given that the relocations were going ahead anyway (see Imrie, Thomas and Marshall, 1995).

Thus, for CBDC, there was nothing to discuss, and, presumably, little to gain by entering into a dialogue about the material impacts of its strategies. In such circumstances, as Bresnan (1988) and others note, there are clear limits to working with (in) organisations. Thus, our response to the specific closures we were encountering, of limited access, data being withheld, even subterfuge, was to develop a range of strategies to try and maintain a relationship with CBDC, while not compromising the specific, critically evaluative stance of our research. In particular, we initiated a series of 'insider' interviews with employees of the UDC through personal and other contacts. Through such contacts we were able to develop a contrasting perspective on the organisation, and, in particular, gain insights into its internal culture and modes of project organisation previously denied to us by some of the more senior personnel we had interviewed. Such covert manoeuvres seemed to be a logical, if not inevitable, step in the research process, foisted upon us by the organisation.

Moreover, the lack of co-operation extended to other parts of central government. For even though quantifiable measures of (output) performance are required by the Welsh Office from CBDC on an annual basis, we were denied access to any of the information. As a Welsh Office officer said to us, 'we're not obliged to give out any of this information, it is strictly between us and CBDC', so commenting on one of the key transformations in contemporary local governance, the intensification of secrecy, non-disclosure, and social closures within the state apparatus. This, though, was an ever-present aspect of the research, and, in part, was related to (organisational) differences of perspective of what evaluative research is. Our commitment to the ESRC required some digging around for causal patterns and relations, placing less emphasis on formal, technocratic, modes of evaluation; or, in CBDC's terms, utilising subjective, or illegitimate, modes of inquiry. Indeed, the absence of recognisable, quantitative, methods in our evaluative frameworks was puzzling to the UDC and our inability to satisfy it about the statistical validity of our information weakened our credibility. The opinions of firms were not seen as statistically valid!

Yet CBDC's withholding of evaluative material has not only been confined to people like ourselves, academic researchers. In interview, Rodri Morgan, Labour MP for Cardiff West, and with a material interest in the impact of Bay developments on his local constituency, alleged that CBDC had obstructed him from obtaining a CBDC-commissioned report related to the effects of the proposed barrage. As Morgan outlined:

> So I took it up with the chairman of CBDC and he said, yes, I am very sorry, I have given instructions to our consultants not to speak to you. So I then went to the Speaker and I said, look this is a matter of privilege, I mean an MP is entitled to any reasonable information, public money is being used to promote the scheme and to do a study and I'm not allowed access to the information in respect of my constituency.

Indeed, as Morgan went on to amplify:

> I thought that was outrageous behaviour on the chairman's behalf. He seemed to have no concept that here I am, funded by the taxpayer, here I am doing something which I recognise myself is going to have a major impact on somebody else, on an MP's constituents, and then not letting him have access to the consultants funded by the taxpayer, even though he is responsible for raising these taxes in the House of Commons every year!

Part of the reluctance of CBDC to 'open-up' to us relates to the closed nature of the organisation, and its own internal culture, methods of working, and knowledge. Indeed, our own experiences were also being paralleled by those of small businesses which CBDC was relocating from parts of Cardiff Bay, suggesting to us that the specific closures we were experiencing were not unique but were part of a wider set of structural relations being forged by the UDC. As one firm commented, 'They're dreadful, unprofessional. I am a thick-skinned businessman who has seen a lot, but I have never met anyone as bad as CBDC', while another firm noted that 'They're a poor organisation with no understanding of industry and wasting public money'. Such sentiments have to be interpreted with caution given that CBDC was embarking on an unpopular course of action, as far as indigenous firms were concerned, which no amount of sensitivity on the part of the UDC could have rectified. One methodological problem we faced was that our own experience of CBDC made it more difficult for us to feel confident in our own ability to evaluate firms' opinions without prejudice against the UDC – a problem we attempted to solve by openly acknowledging it and through group discussions among the three researchers involved in the project.

The end game: (re-)presenting policy evaluation

In the first part of the discussion, we considered some of the manoeuvres being played out between ourselves and the UDC in 'validating' the data provided to us by CBDC. However, a particular, ongoing, concern, relates to questions of control over the presentation and dissemination of research results related to evaluations of the new institutions of local governance. On more than one occasion CBDC has demonstrated an acute sensitivity to academic criticism, and at one stage intimated that it should vet all our writings about it prior to publication. Yet, on the basis of the 'feedback' sessions concerning our fieldnotes, where most things we wrote were construed negatively by the UDC, we have always felt less than confident about permitting the UDC the right to vet our research writings. In part, our attitude has been framed by our experiences, of researching an organisation that time and again has displayed an antipathy to our research, and has failed to open up in all but the most marginal of senses. Why, then, should we permit it a degree of control over the production and dissemination of the research?

However, it would be disingenuous of us to suggest that we have not been affected by the apparent sensitivities of CBDC, and other agencies, to what we have written, or may write. We would hope that it has not stopped us telling the truth as we see it, but it has certainly made us think a little harder about how the truth is phrased and, in particular, what effect publication is likely to have on our relationship with the agencies being studied (see, for example, Morgan, 1972). In particular, there was a real sense in which CBDC felt it knew the answers, and only required the evidence to support its stance. Yet, as Beynon (1988) notes, if research is, in some senses, an inquiry into the unknown, it requires a similar approach in its audience (in this case, CBDC). However, the UDC seemed uninterested in the findings of our evaluative studies and did little to encourage or develop a dialogue, possibly because we were doing little to reinforce its message, of an area 'on the move', of increasing prosperity, with everyone a winner. Indeed, it never contacted us, nor did we receive feedback from it on the reports that we gave it through the course of the project, even though some provided positive accounts of the UDC's programmes.

The irony with all of this relates to the wider stereotype propagated about academics and university research, of the ivory tower populated by out-of-touch researchers. It became clear that CBDC's officers did see us in this way, or, as one of its officers noted, 'What are you really seeking to achieve with all of this, and what will it really do for the people of the city, it's all so impractical?' (see also Beynon, 1988; Bresnan, 1988; Bryman, 1988). Well, we could have said the same about the UDC! Yet, in a context where government is lambasting academics for not trying to relate to 'the real world', the irony is that our attempts to work within it has met with hostility and resistance, and no attempt to reciprocate or develop working relations, that might be described as mutually beneficial, has emerged from the UDC. From our experiences, therefore, working within the 'real world' of CBDC would have amounted to no more than playing the role of 'lap-dog', a means of providing the organisation with academic legitimacy and credibility.

There is also the issue of the reliability of the findings of policy evaluation in an organisational context which is transient, fluid, and prone to high levels of turnover of key informants, bringing with it changing terms of reference and perspectives (see also Bresnan, 1988). Indeed, over the course of the research, CBDC displayed many of the characteristics of a short-term organisation, of volatile switches in policy, a fixation with 'doing', of getting visible results on the ground, and little or no regard to criticisms of its approaches to policy formulation and implementation. This, then, posed a challenge to us in terms of our flexibility, and responsiveness, in reporting such transformations in the organisation and actions of the UDC and, over the course of the research, we have tried to accurately portray, in our writings, an organisation seeking to sensitise itself to the competing demands of local actors.

Conclusions

There is nothing new about social research becoming entangled either in internal organisational or interorganisational, politics. However, we are suggesting that the phenomenon is more likely in urban policy evaluation in the 1990s than it has been for some time precisely because of the changing socio-political contexts within which evaluative research has to take place. Indeed, we were taken aback by the ferocity and frequency of our 'grillings', and have pondered hard about the circumstances which gave rise to them. We cannot discount the possibility that personal inadequacies in our research techniques played a part in somehow provoking them, for, as Bryman (1988) argues, it is important to acknowledge 'the role of personal factors in the facility with which investigators are able to make particular techniques of investigation work for them' (p. 14).

However, we would claim that our own study is merely illustrative of the general trends we highlight in this paper; moreover, we are aware of other urban policy researchers who have received considerably rougher treatment than ourselves as a result of criticising powerful institutions (Loftman and Nevin, 1994).We would argue that at the root of these aggressive reactions is the antipathy to external evaluation of an organisation which typifies key trends in the delivery of urban policy in the 1990s – entrepreneurial, engaged in renegotiating institutional boundaries and responsibilities, and wedded to a 'can do' philosophy which makes a virtue of working 'with the market' and has a tenuous claim to democratic accountability. Such an organisation will not only be suspicious of the findings of independent evaluative studies and the uses to which they may be put by its critics; it will also tend to reject the legitimacy of evaluation by those outside its circle of allies.

The term 'circle of allies' is introduced deliberately. Barnekov et al.'s (1990) review of the evaluation of US urban regeneration initiatives identified a turning away from 'rigorous evaluation' towards 'informal . . . local assessment networks' (p. 29) sharing practical information about what works and what does not. They appear to recommend this approach for the UK. We would argue that extreme care be taken in considering such a recommendation. In the highly charged field of contemporary urban policy, informal assessment networks may easily collapse into circles of allies – i.e. uncritical networks which present a common front in a potentially hostile world. The involvement of researchers within such networks is particularly problematic inasmuch as it may compromise their standing with all parties in live debates about the future of urban policy delivery at the local level.

Acknowledgement

The research on which this chapter is based was undertaken using an Economic and Social Research Council (ESRC) grant, no. R000 233525. We gratefully acknowledge the assistance and support of the ESRC and also the valuable feedback and comments which we received from participants in the ESRC-sponsored seminar on Urban Policy Evaluation, held in Cardiff in September 1993.

PART THREE

Pluralistic Evaluation in Practice

10

EVALUATING COMPETITIVE URBAN POLICY: THE CITY CHALLENGE INITIATIVE

Nick Oatley and Christine Lambert

The City Challenge Initiative

City Challenge was launched by the (then) Environment Secretary Michael Heseltine in May 1991. It was heralded as a significant innovatory approach to urban regeneration and an example of the new institutional framework that is emerging in local governance (Malpass, 1994; Davoudi and Healey, 1994). The government claimed that City Challenge marked a revolution in urban policy and that the stimulus of competition would transform the way in which local authorities and their partners approached the task of urban regeneration (DoE Press Release, 16 July 1992).

The initiative involved local authorities putting together plans for the redevelopment of neighbourhoods which they considered to be of critical importance to the regeneration of their area, in partnership with businesses, the community and the voluntary sector. These plans formed the basis of a bid to government for funding. In the first round of bidding (1991/92) 15 councils were invited to bid. Eleven bids were selected to receive £7.5 million per year for five years. The second round (1992/93) was open to all 57 Urban Priority Areas from which 20 bids were selected for funding. In 1993/94 City Challenge accounted for over a quarter of public expenditure in inner cities (Stewart, 1993), and over the six years 1991–97 it will amount to over £1.1 billion of public expenditure. Although the initiative has been suspended after only two rounds the principle of competitive bidding for urban policy funding has been carried over into the newly established Single Regeneration Budget (SRB) which brings together under one budget 20 existing programmes for regeneration and economic development.

City Challenge has a number of distinctive features. In terms of policy since 1979, City Challenge (and the SRB which supersedes it) represents a marked shift away from the principles that underpin earlier initiatives (Burton and O'Toole, 1993, p. 198). For example, after twelve years of cut-backs and policies which have weakened the role of local authorities, City Challenge identifies a key (albeit enabling) role for local authorities in leading regeneration activities (DoE, 18 February 1992). On the surface, this seems to be a radical departure from the anti-local authority rhetoric and emphasis on property-led regeneration activities that characterised urban policy in the 1980s. However, through the explicit promotion of partnership, local authorities have been encouraged to adopt more corporate practices and inter-agency working and to establish fast-track decision procedures and special sub-committees to expedite City Challenge matters. This has also involved relinquishing overall control to arms-length boards or trusts with representation from business, academia, the community and public agencies.

As well as promoting a change in the way that local authorities relate to their business and resident communities, City Challenge also encourages strategies that attempt to tackle problems in an integrated fashion, linking projects dealing with employment and training, childcare, housing, environmental concerns and crime and safety. Within the proposed strategies great emphasis was placed on attempts to reintegrate disadvantaged areas into the mainstream economy of cities by linking the area targeted for regeneration with the rest of the city.

The most important new dimension of City Challenge was the introduction of a controversial and highly politicised competitive bidding process which was intended to stimulate a fresh, innovatory approach to urban regeneration. Ministers felt that the Urban Programme had become too routine, lacked any cutting edge, and had failed to gain any significant degree of private sector and/or community involvement. Competitive bidding was seen as a way of promoting an entrepreneurial culture in local government and as a way of producing bids which conformed to the government's objectives of creating innovative approaches to economic and social development through partnerships which institutionalised the influence of a wider set of actors, most notably, those in the private sector.

The introduction of competition into the allocation of urban funding and the intention to change the form of urban governance through initiatives such as City Challenge and the Single Regeneration Budget raises a number of important evaluation issues. Great claims have been made about the galvanising effects of competitive bidding in City Challenge in spite of a lack of research based evidence. However, the abandonment of Urban Priority Areas under the SRB and the creation of winners *and* losers through competitive bidding has serious implications for the overall pattern of urban policy funding and the impact on unsuccessful localities. The following sections identify the evaluation issues associated with the introduction of competitive bidding for urban policy and presents findings from research on the impact on local authorities of unsuccessful bidding in City Challenge. The research included a questionnaire survey of all authorities that had bid unsuccessfully and a detailed case study of Bristol, an authority that bid unsuccessfully in both rounds.

Issues in the Evaluation of Competitive Urban Policy Initiatives

The previous section has established a number of distinctive characteristics of the City Challenge initiative: a competitive bidding process that creates both 'winners' and 'losers'; a greater concentration of resources for tackling urban regeneration; a requirement to consider the integration of disadvantaged areas with the mainstream economy of cities; and, significantly, an intention to change the form of urban governance. Its objectives therefore embrace both *substance*, the creation of more integrated and coherent regeneration strategies, and *process*, the formation of effective partnerships with the private and community sectors and the opening up of the policy-making process to a wider range of interests institutionalised through the requirement to set up an 'arm's length' delivery mechanism. This wide-ranging set of objectives raises a number of issues for evaluation.

Most evaluation of City Challenge so far focuses on the measurement of outputs and outcomes in authorities that were successful in winning City Challenge status, though there is acknowledgement of the importance of evaluating the wider process objectives of the programme (see Davoudi and Healey, chapter 11 in this volume, and also Young, 1993). This is particularly important since City Challenge, and its successor, the Single Regeneration Budget, can be seen as part of an approach by central government to further its political aims for local government – in particular, to further weaken the role of local government, to promote an enterprise culture via competitive bidding, and to open up the decision-making process to a wider set of interests. Moreover, in the light of the government's assertions about the benefits of competitive bidding (Michael Howard in *Planning*, 25 September 1992, p. 8) it is important to examine the impact that unsuccessful participation has on localities.

Two questions in particular need addressing. First, to what extent has the City Challenge bidding process influenced the policy process in local authorities, in terms of bringing in new actors, establishing new mechanisms for policy-making and implementation, and shifting the priorities of local elected leadership? Second, what is the impact of failure in the City Challenge competition on the relationships between local authorities and community and business partners and on the ability of local actors to continue to pursue integrated and coherent approaches to urban regeneration?

The first of these questions requires a more qualitative approach to evaluation than the narrow output measurement and value for money orientation of much urban policy evaluation in the 1980s (Gregory and Martin, 1988; Foley, 1992). It ideally requires the 'tracking' of events as they unfold and investigation of the perceptions and views of participants close to the process. Such evaluation also needs to be placed in the context of the specific social and political relationships occurring in localities. Here we would highlight a need to take account of the central-local dimension in the implementation of initiatives such as City Challenge. The history of central-local relations in Britain during the 1980s has been characterised by a good deal of conflict and resistance, focusing particularly on local financial discretion and the introduction of mechanisms that weakened and, in some cases, bypassed local authority powers and controls. Recent accounts of urban policy suggest that relations now are characterised by a greater degree of (perhaps unwilling) consensus and that there is more widespread acceptance by local authorities of new forms of market oriented entrepreneurial planning to encourage private investment and economic development (Stoker and Young, 1993). However, ideological differences over the issues of local democratic accountability and autonomy remain an important area of debate (Stewart and Leach, 1992; Cochrane, 1993).

A second aspect that is likely to influence the response of localities to initiatives like City Challenge is the issue of local institutional and leadership capacity, which is now acknowledged as a significant factor in the ability of cities to respond effectively to external threats or opportunities. Leadership capacity refers to 'the range, stability and durability of local mechanisms and alliances that have been developed that allow a city to respond pro-actively to external economic pressures' (Judd and Parkinson, 1990, p. 21). Research into leadership capacity and urban regeneration has highlighted the variation between cities in the degree of elite coherence, the ideologies embraced by local elites and the degree of co-ordination between business elites and political groups (Judd and Parkinson, 1990). Important factors influencing the capacity to develop coherent and effective responses to economic restructuring include: the degree of political antagonism between elites and political groups; the structure of government and the extent to which authority is concentrated or diffuse; and the limitations imposed by ideology where, for example, local authorities regard the private sector with suspicion or indifference. Part of the agenda of City Challenge has been to encourage localities to develop an institutional capacity favourable to the promotion of private sector investment and economic competitiveness. However, it is apparent that local contingencies, in terms of the history of relationships, the experience of managing urban regeneration and local political priorities, remain important in mediating the outcomes of such initiatives.

The second question concerns the consequences of failure for localities. With only 32 of the 57 Urban Priority Areas securing funding under City Challenge, any evaluation of the initiative (or indeed any other initiative based on competitive bidding) would be partial if it did not address the impact on those authorities who were unsuccessful in the competition. The departure from rational methods of resource allocation based on indicators of need, which played a significant part in the distribution of Urban Programme funding, to a system based on competitive bidding raises a number of issues. First, City Challenge may be viewed by local authorities as an initiative in which a significant number of the participants are set up for failure. It is also understood that losers in the competition are worse off than they were before, due to the top-slicing of resources from other DoE funding programmes. Secondly, the emphasis in City Challenge on the presentation of proposals to government ministers, who were very actively involved in allocation decisions, coupled with the absence of clear decision criteria, opens the process to accusations of political favouritism and patronage. Thirdly, there is the possibility that the allocation of urban funding via competitive bidding will lead to a more inequitable distribution of funding, with some areas consistently winning, while other equally deserving areas are neglected. The awareness of these issues among local authorities has the potential to undermine the perceived legitimacy of competitive bidding and, again, can be expected to influence how local actors respond.

The Impact on the 'Losers' in City Challenge

The information presented in this section is based on questionnaires sent in June 1993 to all those authorities that were unsuccessful in securing funding in either the first or second round of City Challenge (see those authorities marked with an asterisk in Appendix). Twenty-five local authorities out of a total of 57 were unsuccessful in gaining funding from either the first or second rounds of City Challenge. Twenty-four replies to the questionnaire have been received (see Appendix). The intention of the survey was to collect information on the impact the bid preparation process had on the formation of new partnerships; whether it had changed the way in which the local authority operated; whether projects contained in the bid have proceeded in the absence of City Challenge funding; whether there had been a withdrawal of private sector support following the decision; whether the authority was disillusioned with the bidding process; and whether the authority would bid again should there be a third round. The results are shown in Table 10.1. Information was also collected on the cost of participating in the bidding process (see Oatley, 1995).

The survey revealed a wide variation in the cost of bid preparation among the authorities. Some spent as little as £1,500 (Newham 1st round) whilst others spent as much as £345,000 (Islington 2nd round). The average cost of bid preparation from the sample that responded to the questionnaire was £114,080. If this figure is multiplied by 44 (the number of authorities that bid unsuccessfully over the first and second rounds) one arrives at a figure of just over £5 million that was spent by local authorities on unsuccessful participation in City Challenge. The government would argue that this money was not wasted and that new partnerships were formed, new ways of working were encouraged and that some projects are proceeding as a result of the City Challenge bid preparation process.

Indeed, new partnerships were formed as a result of bidding for City Challenge (in 17 out of 24 local authorities), although these new partnerships were dominated by links with the private sector. Whilst some projects had proceeded without the financial support of the initiative in all but one authority, the lack of City Challenge funding meant that aspirations have been scaled down, schemes have proceeded more slowly, and perhaps most importantly the opportunity has been lost to develop a number of integrated schemes that

Table 10.1 Local authority responses

Questions	Yes	No	Nos=24
Have new partnerships been formed as a result of City Challenge?	17	7	24
Have ways of working in the authority changed as a result of City Challenge?	14	10	24
Have projects included in the bid proceeded in the absence of funding?	23	1	24
Has there been a withdrawal of private sector support as a result of the outcome of the bid?	10	14	24
Has your involvement with the City Challenge led to disillusionment with competitive bidding?	17	7	24
If there was a third round of City Challenge would your area participate?	24	–	24

Source: survey.

address the multi-faceted nature of problems in these areas simultaneously. In fact, with one or two exceptions, all of the projects that have proceeded have been supported by alternative public funding, including Urban Programme, Urban Partnership, Estate Action, Task Force, Derelict Land Grant and City Grant. Although some private sector interest could be maintained by securing other forms of public sector funding (in 14 out of 24 authorities) it was not possible to retain the scale of commitment demonstrated in the City Challenge proposals.

In addition to encouraging a more positive role for the private sector and levering private sector resources to support projects, City Challenge also attempted to change the way local authorities operated. Some degree of success appears to have been achieved in this area with 14 out of 24 local authorities adopting more entrepreneurial, more corporate and inter-agency approaches.

Whilst it appears that participation in the competitive bidding process had some positive impacts, a significant number of authorities were disillusioned with it as a way of allocating public resources (17 authorities out of 24). Of the seven authorities that said they were not disillusioned, three were second-round winners, two said they were disillusioned to begin with, and another was Wandsworth, an authority known for its support of Conservative government ideology. Some authorities explicitly stated that the allocation of funds to areas of decline and disadvantage should not be based on competition but on the basis of need. Many local councillors felt bitter at having to subject their submissions – based on demonstrable need – to competition. The perception among local authorities was that City Challenge was based on a highly politicised competitive bidding process which had no objective relationship to levels of need or even ability to deliver (De Groot, 1992, also mentions this). A number of authorities were very cynical about the selection process and felt that there was a lack of clear guidelines about the selection criteria and that decisions were ultimately swayed by political patronage. Many authorities felt their bids were as good, if not better, than some of the bids that were successful and the lack of clear and convincing feedback on the decisions did little to dispel the feeling of disillusionment. In spite of this widespread disillusionment all authorities felt they had little choice but to bid in any further rounds in order to gain regeneration funds for their community.

Bristol's Response to City Challenge

The account presented here draws on interviews with key actors in the City Challenge bidding process together with an examination of official files made available to the researchers by Bristol City Council. Before going on to describe the bidding process in Bristol in 1991 and 1992, some contextual information on the city is presented.

Bristol is a city of approximately 370,000 people and is an important regional centre in the south west. Since the 1970s the traditional industries of manufacturing, tobacco, and paper and printing have declined although during the 1980s aerospace and financial services grew within the economy which helped to give Bristol an image of economic health and prosperity (Boddy, Lovering and Bassett, 1986). It is only since the late 1980s that Bristol's economy has experienced severe difficulties in the wake of defence cutbacks and the effect of the recession on financial services. Many of the other areas invited to take part in City Challenge had been grappling for years with the long-term decline of heavy industries and the problems of economic restructuring. They therefore had much more experience of municipal involvement in economic development and urban regeneration. This relative lack of experience needs to be borne in mind when assessing the city's response to City Challenge.

Despite the general picture of economic health and prosperity during the 1980s there was recognition that parts of the city were experiencing serious problems of unemployment, social stress and poverty (Bristol City Council, 1988) particularly in certain inner city areas (St Pauls/Easton) and on a number of large peripheral local authority estates to the north and south.

In terms of the organisational context it is important to note that Bristol does have a history of conflictual relations with the County, central government and the private sector. Stewart (1990, p. 40) notes that:

> In Bristol the tradition has been one of mistrust and conflict between organisations and conflict over roles and function. City and County have been at loggerheads since the 1974 reorganisation, the private sector fails to speak with a united voice, local government has adopted a conflictual rather than a conciliatory stance towards the new initiatives of the centre. Thus inter-organisational competition is allied to complacency . . . leading to a marked lack of local capacity to respond to the challenges of change in the 1980s.

In this respect Bristol presents a sterner test than some other locations of the capacity of City Challenge to generate close and successful partnerships in policy-making.

City Challenge was launched in May 1991 and authorities were given a tight timetable of six weeks to assemble their bids. The City Challenge invitation was addressed to the City Council and senior city politicians decided that the invitation be accepted, though without great enthusiasm or wide debate. Unlike Bristol, many other authorities gave the initiative high priority (Davoudi and Healey, 1994). The ambivalent response of Bristol's Labour leadership can, according to interviews, be attributed to objections to the principle of competitive bidding and the perceived humiliation of being forced to dance to a Tory government tune. They objected also to the resources being tied not to need but to the demonstration that regeneration would be undertaken in a particular way.

Informal discussions between Bristol City Council staff and officials at the regional office of the Department of the Environment (DoE) leading up to the formal invitation to bid had focused on a major regeneration scheme for the peripheral housing estates of Hartcliffe and Withywood to the south of the city and on the guidance of the regional office of the DoE this area was selected by the city as its City Challenge area. From Bristol's point of view City Challenge seemed to be a development of these discussions, though in practice City Challenge was meant to be wider and more ambitious in scope including highly visible 'flagship' projects, designed to have a clear demonstration effect on urban regeneration, more in keeping with the high political profile given to the initiative by government ministers.

Hartcliffe and Withywood are two large peripheral local authority estates on the southern edge of the city, home to some 25,000 people. The estates were developed in the 1950s and 1960s, a mix of houses and high and low rise flats, to accommodate population displaced by inner city redevelopment. With time, however, a number of serious problems have emerged and the area has acquired an unpopular image. The Wills tobacco factory, a major source of local employment (providing 4,000 jobs) closed in 1989, and as other manufacturing industry declined so unemployment in the area grew, reaching an estimated 13 per cent in January 1992. Approximately 60 per cent of those unemployed have been out of work for over six months (October 1991 figures). In addition serious structural and repair problems affect much of the system-built housing, much of the better housing has been sold under the 'right to buy', and lettings have increasingly been to people with little choice in the housing market. Levels of poverty and ill health are especially high in the area. There are few local community services and the economic and environmental problems of the area are reinforced by isolation from the jobs, shopping and other facilities in the rest of the city. Transport links to the rest of the city and the motorway network are poor. As Bristol's City Challenge bid document stated (Trust in the Community (1992), City Challenge Steering Group Statement), 'here we have a potentially explosive mixture of economic isolation, chronic health problems, poor quality housing and blighted futures', a statement that was to prove prescient with the eruption of disturbances on the estate in the summer of 1992.

The process of assembling the first round bid in 1991 was described as *ad hoc* and rushed. The bid team was led by officers from the Housing Department, who had been involved in the earlier discussions with the DoE. Senior council officers and leading politicians were little involved. There was tacit political support for making a bid in order to protect or enhance resources for the estates, but little enthusiasm for the initiative. Consequently there was no central push for a more corporate response nor the establishment of any political structures to guide and validate decisions. Local ward councillors were keen to attract resources to the estates and were active in steering the bid in acceptable directions, though there was little evidence of commitment and involvement by the political leadership of the council. Communication with the County Council was poor, and relationships at this point were characterised as 'acrimonious and chaotic'. Within the time available, and given the lack of established relationships with the business sector in the city, it proved difficult to involve either the private sector or the community in any meaningful way. The exercise exposed the local authorities' difficulties and inexperience in working co-operatively with each other and with the private sector. The extent of the links that were established in the first round are shown in Figure 10.1.

An important factor that emerged in the first round was the failure to secure any commitment from the Hanson Corporation, owners of the vacant Wills factory site. This was a key potential employment site and needed to be exploited within the broadened focus of the City Challenge initiative. Credible proposals for the site were seen as crucial, but Hanson proved to be particularly distant and difficult to deal with. Proposals for the site were consequently very tentative.

Failure, when it came, was not unexpected. The DoE were very sensitive to both these issues and referred to them in the decision letter (Heseltine, 31

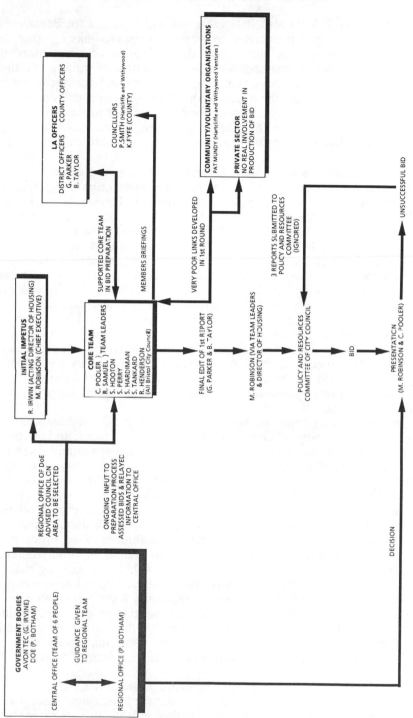

Figure 10.1   Networks in round 1 (1991–92).

July 1993). The DoE letter also stated that the wider impact for Bristol of the bid had not been fully demonstrated, important transport links to other areas needed further development, and greater diversification of tenure needed to be considered. These reasons underline the difficulties of putting together a credible bid acceptable to both central government and the local political leadership. On the one hand the future of the Wills site depended on decisions taken by a multinational business conglomerate, unable to be influenced by a provincial city council. On the other hand some of the reasons (transport links and tenure diversification) covered areas of great political sensitivity in the city.

While various lessons were learned from the first round experience – the need for clear management and leadership and a more corporate response, a requirement to involve a wider range of partners and move beyond a local authority dominated process – there was no effort to sustain relationships or pursue initiatives included in the bid. Consequently when the invitation to bid for City Challenge in 1992 came along the city had to start from scratch and respond within a tight deadline of three months. The preparation process for the second round bid was, however, a much better resourced and centrally supported effort involving a wider range of partners. Again the city was given a strong steer by the DoE towards the Hartcliffe and Withywood estates. While there was some debate in the local authority about whether a bid based on a different area of the city might stand more chance of success, the decision was made to stick with the original area. This partly reflected the active resistance of Hartcliffe and Withywood councillors, a feeling that consistency was important – that the area should not be abandoned – and perhaps also a more strategic view that City Challenge represented an opportunity to exploit the employment potential of the vacant Wills site.

The key differences from the first round bid were the establishment of a full-time team comprising officers from the city and the county, and nominated partners from the private sector, the university and the community, led by a senior officer from the Planning Department. A steering group was set up representing a wide range of interests in the city, including representatives of the private sector, the two councils, Bristol's two universities and the MP for Bristol South. Regular political briefings were held with local ward councillors, and there were much greater efforts to involve the community in the area. Important, symbolically, was the establishment of an office in the area. The involvement of the local community was seen as an important aim of City Challenge. In the first round Bristol was only able to use existing channels and networks to consult the community. In the second round representatives from the community were part of the core team and were represented on the steering group, and the community more broadly were more actively consulted. Although a great deal of work was done to involve the community in the bid stage of City Challenge it was the action planning stage and the establishment of the Trust (the delivery mechanism) where the community's involvement was to be formalised.

The process this time around was therefore characterised by a stronger commitment to corporate working, a more outward looking stance by the city including greater effort to form partnerships with the private sector, wide consultations and discussions and a search for consensus. The response was therefore much more securely grounded in organisational terms and lessons had clearly been learned from the experience of the previous year (see Figure 10.2 which clearly illustrates the contrast with round 1).

However, while it was possible to improve the organisational framework and open the process to a wider range of partners, the production of a credible bid remained politically difficult. The term 'partnership' tends to imply co-operation and consensus, but these cannot be taken for granted in an exercise that brings together people from different organisations, with different motivations and objectives. The negotiation of these tensions was also not helped by the short time period available for assembling bids. For some of the private sector partners in the exercise the generation of development opportunities was a key motivation. For some local authority officers the priority was to improve housing and social conditions on a deprived outer estate. There was evidence of specific conflicts around the issue of housing – whether to apply resources to the redevelopment of some of the existing local authority housing, and achieve the objective of tenure diversification, or to give priority to the improvement and repair of the worst high rise housing blocks. The integration of the area with the rest of the city by the construction of a major road link into the city, incidentally also linking in to a controversial road link constructed by the Bristol Development Corporation, was seen as important by some of the more pragmatic local authority officers, but actively resisted by councillors for whom the proposal was environmentally damaging and a lower priority than improvements to the public transport infrastructure.

The failure to secure the full backing of the Hanson Corporation, which fundamentally undermined the first round bid, remained a significant problem in the second round. Despite a substantial effort on the part of the team to put together a package of proposals, Hanson's support was not forthcoming. As a result the bid document contained rather tentative proposals concerning the development of a business park on the vacant Wills site and a highly qualified statement regarding the construction of the new road link. The rather fragile basis of this central economic proposal was obvious to all of the local participants, and no doubt also to central government. The episode highlights the vulnerability of the partnership approach to regeneration to the wider investment and disinvestment strategies of large global capital, and the particular power that Hanson has over the land and property market in Bristol (Punter, 1993).

The different perspectives of the local ward councillors and the political leadership of the council, which had emerged in the first round bid, continued to be an issue. The political leadership maintained its ambivalence to a central government imposed process and agenda. One of the government's aims was to encourage local authorities to change the way they operate, to adopt imaginative, creative and integrated solutions to urban regeneration, working across departmental and agency boundaries. While key officers from the local authorities embraced the opportunity to work corporately and link up with external agencies, many of the politicians had reservations about this. There was particular resistance to transferring control of the regeneration programme to a separate body at arm's length from the council with minority councillor representation, as the City Challenge rules require. From a political perspective the conceding of democratic control was a price too high to pay. Consequently

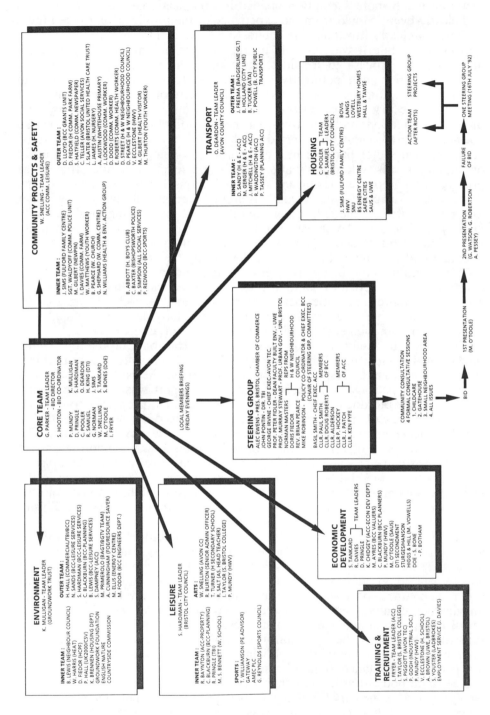

Figure 10.2   Networks in round 2 (1992–93).

it proved difficult to develop detailed proposals on the subsequent management of a successful bid, and interviewees suggested that in the event of success the conflicts surrounding the delivery mechanism were bound to re-emerge.

The more pragmatic and public political stance did not actively obstruct the bid preparation process, but nevertheless there was an absence of visible and committed leadership. Participation in the steering group was sporadic and, significantly, the leading politicians on the council failed to appear on the platform when the second round bid was presented to central government.

At one level therefore the second round of City Challenge in Bristol can be characterised as a more 'successful' attempt on the part of the local authorities to work corporately, to obtain greater involvement from the private sector and to develop a strategy in consultation with a wide set of interests in the city. There is a gap, however, between the serious efforts of the active participants to develop a worthwhile and credible bid, and the less active participants, notably the Hanson Corporation and the leadership of the Bristol Labour group. The position of the leading politicians meant that it continued to be difficult to develop proposals for transport, tenure diversification and the delivery mechanism. The stance adopted by Hanson meant that the bid could not demonstrate effective partnership with a leading player in the Bristol economy nor make definite proposals for the redevelopment of the Wills site.

The decision in the second round to reject the Bristol bid was nevertheless an enormous disappointment for those most closely involved, despite the acknowledged fragility of the Wills proposal. For those adopting a more distant stance on the process, in particular some of the leading politicians, the decision served to confirm their cynicism and distrust of central government, and may even have been welcomed for removing the threat to established power bases and ways of working in the city. One can speculate that rejection twice left the community feeling betrayed – while there is no evidence that the riots which occurred on the Hartcliffe estate in the summer of 1992 in the wake of the decision were in any sense a direct response, they were a dramatic reminder of the sense of frustration felt by some parts of communities in areas such as Hartcliffe.

There are different views on the justification for refusing Bristol City Challenge status in the second round and the 'real' reasons for the decision. Some interviewees emphasised the serious weaknesses of the bid document and the process that has been described, and in that light judged the decision if not justifiable, then understandable. Others felt that other, successful, bids displayed similar weaknesses. A cynical view is that Bristol was being 'punished' for its conflictual stance to central government policies, and there was an unconfirmed rumour that Bristol was removed from the list of second round winners at the last moment when the London Borough of Brent, which changed from Labour to Conservative control in the local elections that year, was substituted. This illustrates the vulnerability of highly politicised competitive bidding processes like City Challenge to accusations of political patronage.

It is possible to add a postscript to these events which sheds an interesting light on the government's comments that all those participating in City Challenge stand to benefit. The specific mechanism of the Steering Group set up for the City Challenge bid process in Bristol has not met since the announcement of the decision, but a small team of local authority officers have been working to keep some of the proposals in the bid alive. This team has argued for the establishment of a permanent area office on the Hartcliffe estate, together with a more general rethink about the need to adopt area regeneration strategies. The City Challenge process, together with the riots, focused attention on an area that was previously marginal in all senses of the word. A number of small projects are proceeding, one with Urban Partnership funding (the Gatehouse Project), and a major private sector leisure project in the area is still under discussion. Less than a year after the second round bid was submitted the Hanson Corporation announced independent plans to demolish the Wills factory and to redevelop the site for light industry and leisure activities.

At a more general level there is a perception that relationships between the public and the private sectors in the city are better than they were, and that county–district relations have improved enormously. Even the city council's political stance in relation to the current central government agenda of partnership and joint ventures in urban regeneration is taking on a more co-operative and pragmatic flavour. However, it is difficult to attribute many, if any, of these changes in the local political and policy environment to the experience of the City Challenge initiative alone. In the 1990s the Bristol economy has begun to display a number of weaknesses that did not exist previously. The contraction of the defence sector and the effects of recession on the financial services sector have made a wide range of interests anxious about the economic future of the city and have led to the establishment of a more coherent business leadership organisation which is linking up with the public sector to promote inward investment (the Western Development Partnership). Furthermore changes have taken place in key personnel in the city council, with the appointment of a number of new senior officers which has contributed to the more pragmatic outlook of the authority.

Conclusions

The theme of this chapter has been that evaluation of competitive urban policy initiatives such as City Challenge requires an approach to evaluation that moves beyond a concern with output measurement and value for money and deals with the substantive and process issues raised by competitive bidding. Two distinctive questions for evaluation are identified: the impact of such initiatives on the process of policy-making and implementation in authorities that participate; and the impact of failure on those authorities who do not succeed in the competition. The survey of authorities who failed to win City Challenge status and the Bristol case study suggest a number of conclusions about these questions.

First, it would seem that City Challenge has been a contributory factor in changing the working practices of local authorities. While it is difficult to interpret some of the survey returns, the evidence suggests a degree of innovation in organisational structures and policy-making processes, and a greater role for the private sector in decision-making. However, the Bristol case study reveals some of the limitations of the City Challenge approach. Convincing partnerships cannot be expected to be established in the short time-scales available for responding to City Challenge. The conflicts of interest inherent in bringing together partners from a range of different backgrounds take time to be resolved, and it is clear that the city lacked the leadership and institutional capacity to demonstrate effective partnerships as well as some other areas.

It cannot be assumed that competitive bidding for urban funding will provoke a uniformly enthusiastic response. As has been shown, in Bristol, the leading politicians, while not actively obstructing the process, maintained a low public profile in relation to the City Challenge initiative and ultimately refused to compromise on certain basic principles. The efforts of some participants to respond to the central government agenda, and to open up decision-making to a wider set of interests, were ultimately constrained by the local political context. Our interviews suggest that the incentive of an additional £37.5 million of public funds for urban regeneration was not enough to override a commitment to local electoral accountability. Moreover, the role of the Hanson Corporation in Bristol's failure to win City Challenge funding is an important reminder of the power of large multinational capital to stand back from initiatives of this kind.

The evidence from the wider set of authorities that failed to win City Challenge status should also make us cautious about the government claims of benefits even in those authorities who fail. On the surface it appears that a significant number of projects proceeded in the absence of City Challenge funding. However, most of these received funding from alternative urban funding sources allocated on the basis of need criteria, and most are oriented towards property and economic development, rather than social and community development. They certainly do not add up to the comprehensive and integrated approach to urban regeneration that City Challenge acknowledges is needed. Furthermore, the extension of competitive bidding to the allocation of urban funding across the board means that the public financial underpinning that is still required to attract private sector investment may not be available on the same basis in the future.

There is also evidence that cynicism and disaffection is a common response to failure and that people at a local level see competition as an invalid way of distributing resources for urban regeneration. There is a strong perception of City Challenge as a 'beauty contest' in which rewards are related to the quality of the presentation, rather than the strength of the underlying case. There are also realistic fears that the extension of the competitive model to other aspects of urban funding will exacerbate divisions in local government, as those who are not well equipped to succeed in competitive bidding experience a downward spiral of funding.

Competitive regimes like City Challenge represent a risky strategy for the government because they rely on incentives that local actors may not respond to and because the competitive element creates losers as well as winners. From the local point of view the explicit departure from rational methods of resource allocation based on need is perceived as invalid. Failure in the competition may actually have negative effects on how people work together – recriminations may set in, cynicism may be confirmed and demoralisation may lead people to conclude that their effort should be directed elsewhere. It cannot be assumed that failure will lead people to work harder together to overcome adversity.

APPENDIX: City Challenge 1991 and 1992: winning and rejected councils, and response to authors' questionnaire

City Challenge 1991

Winning councils

1. Dearne valley * *
2. Bradford
3. Lewisham
4. Liverpool
5. Manchester
6. Middlesbrough
7. Newcastle
8. Nottingham
9. Tower Hamlets
10. Wirral
11. Wolverhampton

Rejected councils

1. Birmingham *
2. Bristol * R
3. Salford *
4. Sheffield *

Uninvited bidders

1. Coventry *
2. Newham *
3. Sandwell * R
4. St Helens * R
5. Stockton *
6. Sunderland * R

Notes:

** Dearne valley was a joint venture between Doncaster, Barnsley and Rotherham.

All winning councils were Labour controlled.

* Authorities which received questionnaire.

R Replied to questionnaire.

City Challenge 1992

Winning councils

1. Barnsley
2. Birmingham
3. Blackburn
4. Bolton
5. Brent
6. Derby
7. Hackney
8. Hartlepool
9. Kensington and Chelsea
10. Kirklees
11. Lambeth
12. Leicester
13. Newham
14. North Tyneside
15. Sandwell
16. Sefton
17. Stockton
18. Sunderland
19. Walsall
20. Wigan

Rejected councils

1. Bradford *
2. Bristol * R
18. Middlesbrough * R
19. Newcastle *

3. Burnley *	20. Nottingham * R
4. Coventry * R	21. Oldham *
5. Doncaster * R	22. Plymouth * R
6. Dudley * R	23. Preston * R
7. Gateshead *	24. Rochdale *
8. Greenwich * R	25. Rotherham *
9. Halton * R	26. St Helens * R
10. Hammersmith and Fulham *	27. Salford *
11. Haringey * R	28. Sheffield * R
12. Hull * R	29. S. Tyneside * R
13. Islington * R	30. Southwark *
14. Knowsley * R	31. Tower Hamlets *
15. Langbaurgh * R	32. Wandsworth * R
16. Leeds * R	33. Wolverhampton *
17. Liverpool *	34. The Wrekin * R

Non-bidders in 1992

Lewisham
Manchester
Wirral

Notes:

*   Authorities which received questionnaire.
R   Replied to questionnaire.

11

CITY CHALLENGE – A SUSTAINABLE MECHANISM OR TEMPORARY GESTURE?

Simin Davoudi and Patsy Healey

Introduction

The City Challenge initiative introduced in England in 1991 has been widely recognised as a significant break with 1980s urban regeneration policy (Parkinson, 1992; Burton and O'Toole, 1993; Stoker and Young, 1993). After a decade of antagonism to local government, it gave a central role to local authorities in establishing machinery for area-based strategies. After a decade of reliance on 'trickle down effects', it represented a rediscovery of 'community' (Lovering, 1991) and of the difficulties those in poor people's neighbourhoods face in getting access to what are often small and untargeted 'trickles' of job opportunities. After a decade of project-based incentives, it called for a five-year strategy and action plan. Evaluating its intentions and impact, however, is extraordinarily difficult. One reason is that its objectives have shifted and evolved over time. A more fundamental problem for evaluators is that the programme targets are multi-faceted and relate to process as well as product. A significant assumption within the initiative is that the mode of local governance is as important as material resources in achieving policy objectives, an assumption which conveniently justifies the relatively small funds available to each partnership. Finally, the initiative has been added to the package of urban policy measures at a time when the institutional relations of local governance are undergoing a major transformation and when the quality of life for many poorer people, particularly where concentrated in particular neighbourhoods, has deteriorated, in terms of prospects for jobs, state social support and family and neighbourhood support (Campbell, 1993).

The City Challenge programmes thus provide a revealing window on the contemporary relations of local governance in England. This chapter focuses in particular on the processes and practices developing around two of these programmes, both in Tyneside. Its specific interest is to explore how far the initiatives are likely to achieve the most challenging of objectives, namely to foster processes which encourage residents of 'areas of concentrated disadvantage' to share in and take control of decisions about the future of their areas and their opportunities. The paper examines whether the modes of governance being invented within the City Challenge initiative have the potential to become routine and widespread within the urban arena and whether in practice they represent moves towards more democratic forms of governance.

The Challenge of City Challenge

The objectives of the City Challenge initiative

The initial aims of the City Challenge initiative combined releasing development potential with targeting to the residents of disadvantaged areas, to improve their access to opportunities, primarily conceived in terms of access to the mainstream economy (DoE, 1991). This was to be achieved partly by strategies which were to focus on routes to jobs and partly by building up institutional mechanisms through which the voice of residents in these communities could be articulated and have an influence on urban governance decisions (Bendick and Egan, 1993).

This institutional emphasis was targeted at both government–community relations and at partnership with the business sector. Local authorities were given the lead in preparing bids and action programmes, but on condition that they brought in partners from the private and voluntary sectors and the local community. The objective was to transform local authority processes, to make them more reflective of business and community needs. By February 1992, these objectives had been reduced to more manageable proportions, with much more emphasis on effective delivery of programmes.

Nevertheless, the emphasis on a three-way partnership remained, along with encouragement to the formation of City Challenge Boards governing an executive team, forming a dedicated agency at 'arm's-length' from local authorities. Central government was to provide flexible funding over the five years, exercising control in relation to its objectives through the initial bidding for programme status, annual approval of action plans, and the machinery of output monitoring. The City Challenge programmes were seen by government to be catalysts, developing mechanisms which could spread to the whole local authority area. This prospect was expected to compensate for the reduction in Urban Programme funding.

However, innovating new processes brings City Challenge programmes in potential conflict not merely with traditional local authority practices, but with central government practices and established relationships between local authorities, their neighbourhoods, pressure groups and business communities. Linking disadvantaged areas and their residents to the 'mainstream' of policy processes and the development of three-way partnerships thus challenges existing patterns of local governance as well as management practices within local government.

The context

This challenge has been set up in a situation where these relationships are in any case under pressure to change for many other reasons. Some of these relate to other institutional innovations produced by the present government in pursuance of its peculiar combination of centralist neo-liberal philosophy (Gamble, 1988; Thornley, 1991). Competencies, both functionally and spatially, have been removed from local government to different public, semi-public and private agencies. Local governments have been forced to reduce budgets while dependence on central government funds has increased. Local authority boundaries are once again being reorganised, and authorities are under pressure to privatise and contract out a wide range of services. Tasks formerly undertaken by government agencies are being off-loaded onto firms and citizens. Decision criteria have been shifted increasingly from universalist conceptions of welfare provision in relation to need, to value for money and the achievement of measurable outputs. These changes have been interpreted as a shift from a managerialist to an entrepreneurial state (Harvey, 1989), or from welfare to workfare (Jessop, 1993). Such conceptions probably overgeneralise a more confused process which currently results not only in the fragmentation of local governance activity, but the disruption of established channels, networks and alliances through which local governments were linked to their citizens and businesses. In this context, the City Challenge initiative could be seen as a new way of co-ordinating the fragments through deliberate alliance formation. Another possibility, however, is that it merely adds another fragment.

But the changes in local governance relations are not solely or even primarily the result of struggles 'within' the state. These changes mediate wider changes in local economies and social life. Economic restructuring has in many conurbations removed sections of industry which once had a voice in local politics, through both owner/managers and workers. The new industries, both the large inward investors and the small business sector, are only uncertainly linked to local governance. In any case, for many such firms, local relationships are relatively unimportant. Meanwhile, patterns of social life are changing dramatically, altering household composition and livelihoods, lifestyles and interest groups (Mingione, 1991).

In the context of these changes, the concepts of incorporation of disadvantaged residents into the 'mainstream' and the empowerment of 'local communities' are deeply ambiguous. The concept of 'the mainstream' economy assumes a model of formal economic relations, in which neighbourhood residents may participate through acquiring jobs or setting up businesses (Bendick and Egan, 1993). Disadvantaged residents are assumed to be excluded, in government discourse, due to their lack of skill and/or motivation, coupled with employer prejudices which could be overcome if business were more involved. More contact between business and the community would in turn create role models to encourage residents to seek entry to the formal economy. However, the scale of employment opportunities in the formal economy is not expanding overall. If residents from City Challenge neighbourhoods increase their access to such jobs, their gain will be someone else's loss. Many poorer people have to find other survival strategies than dependence on the formal economy. This may include dependence on state support, but could include informal economic activity, reliance on family networks, 'going on the black' or participation in criminal economic networks. A critical issue affecting people in such situations is whether policy initiatives reinforce or limit the range of choice of survival strategies for their households.

The concept of the 'mainstream' polity may be seen to refer to the institutions and relations of formalised governance. Disadvantaged residents are seen, in government statements, to be excluded from participation and making claims for their interests by a combination of their lack of skill in organising and representing their views, their lack of motivation ('apathy') and the impenetrability (lack of responsiveness) of local authority politics and bureaucracy. The various 'community professionals' built up through past rounds of policy initiatives in areas of concentrated disadvantage are now often presented as part of the problem, mediating on their own terms the relations between residents and government. Incorporation of residents into 'mainstream politics' is thus now seen to involve direct representation by residents. Yet this mainstream politics is itself changing, due to the internal changes within local government, and the different demands of contemporary businesses and citizens. As a result, old networks are disintegrating, new ones forming, focused on new arenas and different forms of representation (Stoker and Young, 1993).

A consequence of these economic and political changes is that the concept of a 'local community' is even more problematic than in the post-war period. Coincidence of interests between firms in a place, residents in a place and governance mechanisms in a place does not exist as a result of some geographical logic. If it exists at all, it has to be actively negotiated among those with a 'stake' in an area. In many urban neighbourhoods, households share little else than their common residence.

Thus, the political challenge for City Challenge is to innovate, through the three-way partnership, consensus-forming processes across disrupted relations between businesses, citizens, and governance machinery in cities, where the crisis of representation and political involvement in 'areas of concentrated disadvantage' is part of the wider contemporary crisis of disjunction between economic and social life and political organisation. As Mayer (1992; 1993) argues, these crises create opportunities for innovation at the local level. This is one reason why the City Challenge initiative has been welcomed enthusiastically in some areas (e.g. De Groot, 1992). But they are also full of contradictions, while the instability and fragmentation of relations and networks allows powerful groups to impose their criteria and practices discretely and often unchallenged.

Economic and social life in 'areas of concentrated disadvantage'

Such areas, primarily poor people's neighbourhoods, are literally at the uncomfortable sharp end of these changes. These neighbourhoods are places where individuals and households struggle to survive in conditions where patterns of economic support, of state support and of family and household support and opportunity are not available to them in forms familiar to their parents' generation or in their own childhoods. It has always been the case in cities that poor people and those 'disadvantaged' in relation to dominant relations of a society cluster in neighbourhoods to which they can get access (as a result of forms of tenure, low standards and/or low costs), and that the conditions of people both within such neighbourhoods and between neighbourhoods are highly varied. Yet in the post-war period, many such neighbourhoods, in both old and new housing environments, were places where people could see opportunities increasing, in terms of jobs, health and education. What is different today is the massive extension of those in unstable and informal economic situations. This is now understood as more than a cyclical phenomenon, the existence of a 'reserve army of labour' in times of recession, to be absorbed in times of growth, as the Community Development Project (CDP) analysis argued in the 1970s. It is a structural consequence of the changes in the mainstream economy.

'Areas of concentrated disadvantage' may thus be understood as places where those actively marginalised by dominant economic and political relations and cultural mores tend to concentrate because here they find space to exist, by choice or lack of it. Once a sufficient concentration of such households has developed in a neighbourhood, the labelling effects will adversely affect others living there. Yet the diversity of households within such neighbourhoods may be very great, and the tensions within them as substantial, and more significant for daily life, than between such places and others which do not carry labels of marginalisation. Faced with these circumstances, households develop all sorts of strategies for survival, creating multiple forms of economic relation. As Mingione (1991) argues, these multiple relations are typically associated with the reinforcement of reciprocal relations, based on family, kin and neighbours, rather than associational relations based on workplace, party and interest group. Changes in public policy have in any case diminished the range of choices available to residents, in parallel to the loss of economic opportunity in the formal economy. In contrast, the 'criminal economy' has been expanding vigorously, creating both opportunities and threats for neighbourhood residents. This means that in such neighbourhoods, daily life is lived according to different norms and cultures than typically among those absorbed in 'mainstream' economic and political relations. If citizens are to learn skills to enter the workplace and/or engage in local governance, how do they integrate these with the norms and cultures necessary for their survival in a range of informal and reciprocal relations? There is anecdotal evidence that the community leaders who work with government initiatives may develop their own capacities, but that this gives them access to opportunities which enable them to leave the area. For others, as Mingione (1991) suggests, it may be more rational not to participate in the 'mainstream', unless what is offered is sufficiently attractive primarily in financial terms. The implication is that there are likely to be deep divisions among residents and within households in such neighbourhoods about how and how far to 'collaborate' with 'mainstream' urban policy initiatives.

In this context, the challenge of the City Challenge initiative can in part be seen as a welcome move to 'break down the labels' as used by the 'mainstream' economic and political forces (such as the insidious power of postcode labelling), an exercise in 'mutual learning' (Friedmann, 1987), to bring business leaders and politicians in direct relation with citizens in 'areas of concentrated disadvantage' and encourage them to see the world through the eyes of the 'other' while providing knowledge and role models for residents. It could thus increase the choices available to neighbourhood residents. But it carries with it a more disturbing pressure by the representatives of the 'mainstream', to require conformance to their norms and cultures (Mackintosh, 1992). The City Challenge initiative thus has the potential to be used as a project of cultural transformation by 'mainstream' élites faced with divergent survival behaviour. The citizens of such neighbourhoods may perceive government policy initiatives as potentially destructive of their interests and survival opportunities. This may be an important factor in their reaction to such initiatives. The micropolitics of who, within a neighbourhood, 'represents' the 'community' and how this relates to the social and economic relations of others in the same neighbourhood is therefore a key issue in evaluating the effect of a policy initiative. So is the nature of the culture and style which evolves around each City Challenge programme. Through observing both the micropolitics and the style of the initiative, we can examine whether the emerging policy processes are effectively empowering residents through absorbing their own ideas understood in relation to their norms and values, or whether such processes are merely yet another form of 'co-option' and control by the dominant 'mainstream' groups.

An approach to evaluation

Evaluating an initiative such as City Challenge is extraordinarily difficult, not only because its objectives are wide-ranging and its focus unstable and diffuse. Its 'targets' primarily focus on making relationships, for example, establishing links between training places provided and routes to jobs, or active involvement in developing and managing projects in order to build the confidence to seek training or a role in formal politics. This suggests that policy evaluation needs to take a qualitative and multi-perspective approach, preferably over a range of time-scales. The tension between the government's narrow output monitoring regime for City Challenge projects and its encouragement of local evaluation teams looks set to head into a confrontation on appropriate evaluation approaches.

Another approach is to stand back from evaluation in terms of the goals of participants, to explore the contribution of the programme to wider objectives – to economic development, the widening of social opportunity and/or political empowerment. This requires specification of a critical stance from which to set up evaluation criteria and methodology. In this paper, we focus on the issue of political empowerment. The City Challenge initiative is directed, in theory, at transforming local governance relations to make them more responsive to business interests and citizens' demands. Is this just a rhetoric, masking the reinforcement of central control and/or continuing neglect of the reality of life in poor people's neighbourhoods, a form of legitimation device? Is it just a temporary gesture, to give the appearance of action? Or is the approach likely to have, as it claims to attempt, significant effects on the processes of neighbourhood governance? If so, are these likely to increase the opportunities for participation by neighbourhood residents in the governance of their neighbourhoods? If so, which residents and whose interests are likely to benefit? This focuses attention on the power relations of involvement in City Challenge programmes. Drawing on discussion by Fischer (1990) and Drysek (1990), we can ask whether the participation pattern and mode represents a tendency towards technocorporatist practices, allowing domination by political, administrative, professional and business élites in some form, or whether it is possible for residents to obtain real leverage over agenda setting and project delivery, and what are the conditions which encourage this (Forester, 1993; Healey, 1992; Hillier, 1993; Friedmann, 1992). Put another way, can residents change the style of 'mainstream' governance or must they conform to it in order to participate?

To research these issues requires specification of the meaning of governance, and of what constitutes moves towards more technocratic or more participatory processes. The term 'governance' is being used in Britain to convey the range of 'service' delivery mechanisms and regulatory systems which now exist to devise and implement policies. It expresses the shift from provision by formal local and central government structures to the contemporary fragmentation of agencies, and of responsibilities between public, business, voluntary and household/kin/friendship spheres.

The forms of governance processes may vary enormously, being complex outcomes of structural constraint and specific contingencies (Healey and Barrett, 1990). To identify the constraining or structuring tendencies in such locally specific evolving forms, one approach is to develop criteria with which to explore such forms in particular instances. A way forward is to combine the insights of the political economy of interest representation and mediation, with its focus on interests, stakes, strategies and networks, and who gains and loses (Marsden et al., 1993; Healey et al., 1988), with the discussion of the micropolitics of social interaction (discourses, language forms, organisation of arenas of interaction) through which issues are brought forward, filtered and consolidated into strategies and action possibilities (Forester, 1989; Healey and Gilroy, 1990; Myerson and Rydin, 1991; Bryson and Crosby, 1989; Offe, 1977). Such an approach to analysis is now being labelled as institutionalist (Rydin, 1993, Healey et al., 1995).

This suggests that an evaluation of the potential for political empowerment and tendencies towards, rather than away from, participatory forms of governance should focus on the following:

(1) Actors, strategies and networks
Who gets involved in neighbourhood governance activity, how do they become involved (in relation to roles, arenas and social networks), how does such involvement relate to their strategies, what interests and values do they bring to governance work, and how do they relate to all those with a 'stake' in a neighbourhood?

(2) Discourses and arenas
How are agendas established, resources, rules and ideas identified, acquired developed and deployed; through what discourses (systems of meaning) and styles (modes of expression), and with what impact in terms of filtering the potential range of issues and systems of meaning and expression among those involved; and among all those with a 'stake' in a neighbourhood?

Those governance processes which tend to widen the range of those involved, their interests and discursive forms, can then be interpreted as moving towards more participatory democratic forms. Those which narrow the range could be seen as technocorporate if the narrowing focuses on professionals, administrators and business élites. But the narrowing could take other directions, as in paternalist forms of traditional labourist/unionist politics (Gyford, 1985) or the clientelistic forms characteristic of Italian politics (Mingione, 1991; Eisenstadt and Lamarchand, 1981).

The rhetoric of City Challenge certainly presents it as a widening initiative, which accounts in part for the welcome that it has received. But in assessing the governance practices through which the various City Challenge programmes are evolving, it is important to identify the dynamic interplay between constraint and innovation, or to put it more theoretically, the relation between structure and agency. A sensitive analysis of governance processes should be able to identify what is 'framing' or structuring the processes and what is contingent 'invention' within these framing constraints. At the end of analysis, it must be possible to answer the question: who controls the processes, and how enduring are their forms? If those who do the work control the process, and if these in turn attempt to bring forward (or 'represent') the interests and values of those with a potential stake in an area, in ways which are seen to be fair and knowledgeable by these people, and if the processes evolved look set to endure, then a City Challenge programme can be judged as moving towards sustainable democratic participatory forms. The extraordinary difficulty of achieving this in the contemporary conditions of many poor neighbourhoods has already been highlighted.

Policy Processes in Two City Challenge Programmes

In this part, we apply this approach to an analysis of the early stages of two City Challenge programmes in the north-east. The Newcastle City Challenge programme was a 'pacemaker' case, submitting its Action Plan in 1991, and entering its first year in April 1992. The North Tyneside programme followed a year later. Our interpretation draws on multiple sources of evidence[1] and is contestable. However, the balance of evidence supports our account. We focus on two arenas of interaction, the preparation of action plans, i.e. the agenda of projects in a programme, and the evolution of the formal structure of partnership, the board, or equivalent. It should be stressed, however, that our account does not assess the subsequent dynamics of the partnerships.

Newcastle West End Partnership

Newcastle City Council has a long history of community development work in its disadvantaged areas. The West End is a large area, with 35,000 people, including the wards of Benwell, Elswick, West City, Scotswood and Moorside. It was the location of major urban renewal work in the 1960s, and saw a shift from redevelopment to renovation in the late 1960s. In the early 1970s, Benwell was the site of one of the nationally funded Community Development Programmes. Many of those involved are still associated with the area in some form. In the late 1970s, it was part of the Newcastle-Gateshead Inner City Partnership funded under the government's Urban Programme. This led to the creation of Priority Area Teams which are still in existence. As a result of this history and the links with residents over housing projects during the 1970s and the 1980s, the area is relatively rich in networks linking the City Council to community development activity. There were also groups which had evolved within the neighbourhoods to make demands on the City Council and other agencies, notably the Scotswood Area Strategy Group. In Scotswood, prior to City Challenge, experiments were underway in partnership between residents and the council in which the community was developing a powerful voice. This could be seen as one precursor of the possibilities the City Challenge programme was to release.

The second precursor was an initiative fostered by the business community. The Newcastle Initiative (TNI), the first business leadership team sponsored by the Business in the Community initiative, undertook a community development programme in one neighbourhood, Cruddas Park. Relations between the local authority and the business community had been pragmatic and reasonably comfortable over the years. The city council had worked in partnership with the private sector on a number of schemes, notably a large city centre retail mall, Eldon Square. The TNI provided a new arena for articulating business interests and relations with the community. In principle, therefore, the Newcastle West End had the organisational elements for the kind of three-way partnership envisaged by government. In addition, the council had a well-established capability to mobilise to respond to new opportunities created by government policy initiatives.

At the same time, government was sometimes critical of the way the local authority operated. It was said that Urban Programme projects were not always implemented to timetable. While the authority had strong central leadership at the political level, it remained departmental in its operations, leading to co-ordination problems. Ward politics remained traditional, focused on competition to capture resources for neighbourhoods. Councillor relations with constituents tended to the somewhat paternalist forms of the old labourist/union political nexus which has long dominated north-east politics. The West End itself had a growing representation from a newer generation of councillors who promoted a 'new left' agenda of issues and processes (Stoker, 1988), but these found it hard to get leverage on council politics generally. Local community groups got access to council resources typically in a mediated form, through the professionals, either within the council departments, or the various voluntary agencies (it was this process that the Scotswood campaigners sought to change). The City Challenge initiative therefore came with a price. The local authority was 'in charge' of developing the initiative. But it was under pressure to show that it was internally co-ordinated and had both a stronger and more strategic relation to the business sector and greater direct representation by people from the local community, in an organisational structure at 'arm's length' from the city council.

Organising an action programme

The council's original bid proposed a board, with equal representation from the city council, the community and the business sector. This took time to set up. A 'shadow' board handpicked by the city council was set up in late 1991. This held useful discussions but appears to have had little impact on the action plan. The bid and the 1991 action plan were primarily prepared by local authority officers, with the Development and Housing departments taking the lead. Given the tight timetable set by central government, there was little alternative, but this in itself highlighted the lack of existing arenas for strategic discussion with either the business sector or the neighbourhoods. Officers and politicians also saw themselves as engaged in a political struggle to capture funds for the city by winning the competition, while at the same time maximising their capacity to control agendas. The 'Action Plan 1991' was dominated by housing projects. This reflected the council's perception of the issues in terms of a housing problem. (In contrast, community agendas stressed unemployment, social disintegration and crime.) The Education Department added a project to build a new school, linked to multiple community services, to replace Redeswood school on the margins of the area.

Nevertheless, structures were created at this stage to articulate the community and business view. Discussions in late 1991 showed a move to greater interdepartmentalism, linking housing, economic development and educational objectives together around individual projects. There were genuine problems in achieving this, as several departments, notably education, were themselves undergoing reorganisation due to changes in government funding and policy. Some projects promoted by the community and business groups were included in the action plan. But the shadow board was obviously managed by the politicians and officers. Meetings were organised through the committee section of the Chief Executive's department, with large bundles of papers in a local authority format. These meetings were held in public, but were carefully prepared with pre-meetings and officer briefings to councillors. The business and neighbourhood representatives found these arenas alien, in their scale (often 50 or more people) and their style (the form of the papers and the discussions).

A City Challenge director, a former city council officer with a background in the Audit Commission, was not appointed full-time until April 1992, and it took some time before an effective and separate executive office for the West End board could be established. For a time, this was housed in the city council offices, premises in the West End being available only from autumn 1992. The Board was formalised in January 1992. But it continued to operate in the manner established by the shadow board. It is chaired by the leader, or deputy leader, of the council.

However, slow shifts in practice developed during 1992. Both the community and business representatives began to hold pre-meetings, paralleling local authority practice, with the result that board meetings became a form of 'three-way caucus'. A community facilitator was appointed to help the neighbourhood representatives articulate views. As these processes evolved, difficult decisions had to be faced about the content of the programme. The original bid was an amalgam of projects put together primarily to win the competition. The action plan was more realistic, but still contained projects which had not been adequately appraised before inclusion or which proved difficult to realise with changes in the political and economic context. With real decisions about resources to be made, and a stronger sense of how to influence the local authority nexus, business and neighbourhood members on the board began to challenge decisions made previously and to call for new priorities. Thus the mechanism of the initiative served to release a new dynamic in relations among the three 'partners'. By 1993, there were, however, still complaints about the style of meetings. Business members also complained about the lack of direction and focus in the programme, as well as the form of meetings and the style of paperwork. As relationships evolved, the community voice became stronger in programme development, and the business voice less influential. There were also questions about representation. The business sector was represented primarily through a new nexus of business leaders and agency professionals which emerged in the conurbation around the economic and political changes of the 1980s. Community representatives were often challenged by other sections of community interest, with involvement sometimes causing divisions at the heart of households.

Analytical comment

In terms of actors and arenas, a shift can be seen from the traditional practices of a big city Labour authority, towards an arena where more voices are expressed and heard. As the City Challenge initiative evolved, central government, business and community representatives have challenged a strong labourist/trades union tradition, with overtones of paternalist clientelism, serviced by highly competent officers within a departmentalised structure. However, the business voice was primarily that of the dominant business/agency grouping built up around government policy initiatives in the 1980s. It tended to exclude both a wide range of small firms, and some of the major companies. However, involvement provided a vehicle through which this particular business nexus has been able to promote its interests in the area and protect its interests in other parts of the conurbation. Where business representatives have no specific interest, they are unlikely to see the point of involvement in what appear to them to be laborious and inefficient processes. Only a few saw the role of benevolent patron as commercially useful. The neighbourhood voice was expressed largely through pre-existing community organisations. Their relations with other neighbourhood residents was often problematic. Central government politicians and civil servants loom behind all these, as they did before, in their role as funders, both of the programme as a whole and in the approval of specific projects. Having approved the bid (through the competitive process), the DoE retained control through requirements for technical procedures in project appraisal and monitoring. Effective performance in these routines, combined with annual approval of the roll-forward of funding, gave central government a powerful tool in shaping the programme. To conclude, while the balance of power in representation was certainly wider than in the preceding Urban Programme arrangements, and the local authority had less control over spending priorities than at the start of the programme, Newcastle West End City Challenge in its early days was hardly a three-way partnership on equal terms.

As regards the operating procedures of the board and the management of projects, these remained dominated by local authority styles. The other partners had to learn these styles or be submerged. Both the business and community representatives then found sufficient voice to challenge the agenda of projects and, to an extent, the organisational style. It is difficult at this stage to assess whether innovation is occurring in the practices of any of the participants, although some agencies claimed that their own organisations had changed to give more priority to actions in the City Challenge area (e.g. Tyneside TEC).

It seems likely that the main changes relate to technique (project management) and priorities rather than to forms of representation, i.e. to forms of local governance. There has been little sense of an active strategic debate among all those with a 'stake' in the area on problems and possible actions.

A realistic hypothesis as to who 'controlled' the City Challenge processes in the early days in Newcastle would suggest that the DoE played a major role, through the project appraisal and annual review procedures. It was this pressure which pushed the local authority to give more attention to the voice and demands of business and community representatives. This in turn provided the opportunity for major private investors to negotiate subsidies for their projects, although it is likely that this would have happened anyway, if less publicly. In this context, the community voice has acted to expand and shift established agendas, capture small funding for specific projects and bring issues into the arena of public debate. A possible interpretation is that central government controlled the resource flows through defining the rules of access to them. The local authority then sought to capture control through the rules governing discussion practices and through the agenda of projects. Within this process, both business and community representatives had some impact in modifying both the local authority rules and the agenda of projects.

The Newcastle City Challenge programme was thus at the sharp end of the contemporary struggle between central and local government, and between formal government, the business sector and citizens, for control over governance form. Given Newcastle City Council's traditions of working with community representatives, and the extent of neighbourhood organisation, there were genuine opportunities for widening the power base and moving towards more empowering policy practices. Experience in 1993 suggested that this widening dynamic was continuing. However, these shifts are held in check by the technical management procedures set in place by central government and by the weight of traditional local authority practices. It is therefore not clear whether the democratic potential of City Challenge can be realised in this situation, or even whether the more technical criteria of better integration, responsiveness and project delivery emphasised by the DoE can be achieved more effectively than through the Urban Programme mechanism. It remains an open question whether the City Challenge mechanism as it has evolved in Newcastle is sustainable and desirable in itself and/or as a model for neighbourhood governance elsewhere in the city.

North Tyneside City Challenge

North Tyneside too has a history of urban policy initiatives. One of the key council officers developing the City Challenge initiative worked in the North Tyneside CDP. However, North Tyneside Council was not invited to bid for pacemaker status under the City Challenge initiative because in the late 1980s it had developed a reputation for inefficiency, conflictual policies and remoteness from popular demands. Links with business were slight. The council, and local action groups, tended to take oppositional stances, to each other and to agencies such as the Tyne and Wear Urban Development Corporation (TWDC). This reputation was dramatically expressed when, after the riots in Meadowell, an estate now within the City Challenge area, Michael Heseltine declared that the council would not get City Challenge funding unless it put its house in order. It was thus catapulted into the national political limelight.

The council was already beginning a very substantial transformation, led by a new cadre of senior council officers, with the direction and strong support of key councillors. The rationale for this was primarily to effect substantial savings and improve the Council's reputation in central government's eyes, in view of its dependence on central government funding, including for City Challenge. There was also a range of community groups in Meadowell, developing around initiatives to improve conditions, and a community forum in one of the other neighbourhoods. But the area lacked the richness of groups and links with the local authority which existed in Newcastle's West End.

Organising an action programme

Obtaining City Challenge was given a high priority within the council, with the emphasis on demonstrating that new relationships were already in place and capable of developing and delivering an action plan. With this in mind, business and agency representatives were drawn into the process of formulating the bid. The City Challenge area has a population of 36,000, and includes a substantial commercial and industrial zone along the A19 corridor, including existing firms and development sites. There are four neighbourhoods, Meadow Well, Howden, Wallsend and North Shields. Community representatives from these four areas were also involved in bid development. These representatives were elected during public meetings organised by council officers. The bid preparation team could thus be presented as broadly based. It also had the benefit of the experience of the first round of City Challenge programmes.

The proposed partnership board for the initiative reflected this participation, including a range of existing *ad hoc* groupings. Those involved became the members of the board as finally constituted. The groups were seen to represent the business sector, the community, economic development (including TWDC and Tyneside TEC), housing issues and the council itself. Council representation was a mixture of officers and councillors. The board's Chief Executive, appointed in late 1992, was a council officer.

Following approval of the bid, the board was formalised with the appointment of a chair from the business sector who has development interests in the area. Agency representatives who were not formally on the board were invited to meetings, but these were not open to the public. The style was very deliberately that of a business board, combining strong executive leadership with a free-ranging discussion of issues, at least initially. Much of the work was undertaken by a core executive group, preparing material for the main board meetings. This consisted of the board chair, a key council officer, and four board members (from the community, the TWDC office and others from the private sector). In board meetings, the language and mode of expression were those with which the business and agency representatives were comfortable. The neighbourhood representatives found it hard to contribute in what for them was a daunting arena, particularly with regard to the executive group. The dominance of business and agency interests and practices was in sharp contrast to Newcastle City Challenge. This style was reinforced by a management training course for those involved in City Challenge work. However, most of the neighbourhood members of the board did not attend the course, finding the environment alien. The style partly reflected North Tyneside Council's deliberate attempt to mimic a style it believed central government preferred. But it also reflected the particular geography of the North Tyneside City Challenge area. A number of businesses had a significant stake in the area, and many of the projects in the action plan were for business and property development. One consequence was that board meetings were punctuated by formal statements of 'interests' in particular funding proposals. However, this business 'voice' did not necessarily represent all business interests with a stake in the area, and some private sector members felt a little on the edge of things.

The community representatives, among whom are the only women on the board, had a hard time in this arena. Their organisational base was less well-developed than in Newcastle. They were not initially provided with support resources, and remained largely silent in board meetings. They had little input into the discussion on the definition of the main agenda of projects. They were clearly overwhelmed by the world of business and local government. In 1993, however, a community facilitator was appointed. This may help the community representatives find a 'voice' in board meetings.

Despite the appearance of dominance by the conurbation's business/agency nexus, the local authority retained a strong grip on the programme, primarily through its officers. The strategic agenda was set by senior local authority officers, with the support of senior councillors. The procedures of the board and the energetic adoption of new management procedures for City Challenge projects reflected the council's decision to transform itself and signal a new style of governance within North Tyneside. The objective was clearly to capture central government funding more effectively. The community must expect to benefit through the attraction of funding to the area generally, and by more effective 'trickle down' of benefits from projects. One consequence of this strongly managed approach was that much of the agenda of projects was already determined early on, so board members found themselves discussing the detail of project implementation rather than strategy. There is some concern that member interest in board affairs was being lost as a result.

Analytical comment

For North Tyneside, City Challenge provided the opportunity to build up a new alliance between the local authority and the Tyneside business-agency nexus. In contrast to Newcastle, such links barely existed previously. Through this, councillors and officers hoped to maximise their chance of 'capturing' government transfer payments for the area at a time of severe financial cutbacks and to improve their capability to deliver programmes for their communities. The business-agency nexus was actively interested in the programme, both to promote property and business development initiatives and to protect their interests elsewhere. The practices evident in the preparation of the bid and the action plan reinforce the contrast with traditional local authority practices. Board meetings were set up to mimic private sector practices, or rather those in companies of a particular scale and type, and the dominant discourse was a mixture of business and government terminology. The organisational style was thus deliberately technocorporatist. One result was that the community representatives had little impact on the programme. The dominant actors had a strong commitment to benefiting local people, but their approach was essentially paternalist, developing the action plan on behalf of local people, rather than actively with them. (In contrast, some community members in the Newcastle City Challenge programme had become active and influential participants.) This approach may bring greater opportunities for neighbourhood residents, but does little to incorporate them in the mainstream of political life in North Tyneside.

The potential tension between the local authority and the DoE in North Tyneside is suppressed through this approach. In effect, the local authority is seeking to maximise its opportunity for control by deliberate conformance with DoE demands and collaboration with business interests. Technocorporatism as a style has been chosen as a strategy to maximise returns from subsidy to the area. It could be seen as a way of incorporating the local authority itself back into the 'mainstream' of central-local politics.

Given the parallel general reorganisation of North Tyneside District, there are also opportunities for the strategic alliance and management practices built up around the City Challenge programme to be extended to other areas of local authority work. In this sense, the North Tyneside City Challenge processes may become more sustainable and more extendable than in Newcastle City Challenge. But they are not, so far, more democratic.

Conclusion

Government objectives for City Challenge seek 'successful partnerships' with all those with a stake in an area, consolidated into a three-way partnership between the local authority, business and the community. In this way, the community and business are to be integrated into the 'mainstream' of local governance and contribute to the government's wider project of transforming local governance processes, to make these more responsive to the expressed needs of business and the community. From the government's point of view, the North Tyneside City Challenge is possibly more successful than the Newcastle City Challenge, whose progress has been more halting and dominated by traditional local authority practices. Yet the neighbourhood residents involved have so far had presence but little voice in North Tyneside, while in Newcastle, their voice has made an impact. In North Tyneside, a new strategic alliance was constructed, producing an effective partnership between the local authority and the Tyneside business-agency nexus, watched over by the DoE. In Newcastle, uncomfortable co-existence rather than partnership was achieved between the local authority and community representatives, with the business-agency nexus expressing frustration and the DoE, concern.

The difference between these two experiences reflects many factors, among which are the differences between the areas, the North Tyneside City Challenge having many more larger firms within its boundaries than in Newcastle, where the companies are on the periphery or beyond the boundaries of the area. As a result, the Newcastle programme has a stronger emphasis on housing and community-based initiatives. Newcastle City Council also has a long tradition of community development work in these neighbourhoods. Yet the difference also arises from local authority strategy. North Tyneside sought to maximise political and financial returns by deliberately embracing a transformation of style. Newcastle sought to maintain its political networks and increase community access to them, implicitly confronting government.

The power of central government, through providing the resources and controlling the rules of the programme, remains very strong. Its rules and objectives, despite the rhetoric, emphasise commercial approaches to organisation and management and are influenced by Treasury controls. These procedures are deeply technocratic. As a result, while ensuring community involvement in the partnership, central government pressures in themselves are unlikely to lead to active community participation. Any benefits they bring to the community are likely to be indirect, through greater efficiency and targeting in project delivery and more transparency, encouraged by the monitoring regimes. However, these regimes represent a technology which is difficult for neighbourhood residents to relate to. It requires a commitment to rational procedures which may be very alien to the cultures and processes upon which their survival depends. Central government thus achieves its dominance not only through control of resources, but through control of the rules, and the discourses which go with these.

If central government encourages community presence rather than voice in City Challenge partnerships, then pressures to counteract technocorporatist processes have to come from elsewhere. The options are either the local authority or neighbourhood residents themselves. Newcastle City Council clearly had a philosophy and a neighbourhood politics which encouraged more involvement. Building on this and the encouragement provided by the initiative itself, community representatives have expanded their influence on the content and style of the programme. Yet the neighbourhood representatives had a hard task learning the local authority style and confronting local authority positions. It is still the case, as the CDP reports recorded in the 1970s, that local authority bureaucracy was part of the problem for neighbourhood residents (Edwards and Batley, 1978).

Thus in neither of the City Challenge programmes is there much prospect of a movement towards participatory democracy. Neighbourhood residents have very limited control over programme content or policy processes, even though some of them are actively involved in project delivery. Nor is it clear how the involved residents relate to all the rest. To move local governance processes forwards towards more participatory forms, however, requires more attention, not just to the politics of interests and influence (who is involved), but to the manner of their involvement and the style in which matters are discussed, decided and acted upon. Community participants often remarked on their sense of 'discursive marginalisation', but the significance of this was rarely understood by the other participants. In effect, a key dimension of cultural difference remained largely invisible to them. The broad mass of residents, often disconnected themselves from the community representatives, were left to gain whatever material benefits may arise as a result of programme investment – more job choice, better skills, more housing choice, less crime. In a context where the prospects for real democratic involvement were limited, as in these cases, City Challenge programmes which maximise these material and social benefits in the short term have merit. Technocorporate practices may be the most effective means to such an end. However, in the longer term, the costs of alienation from 'mainstream politics' could be considerable. Thus the agenda of future urban policy should continue to attend carefully to both process and material products.

North Tyneside City Challenge perhaps provides an example of attempts to incorporate more residents into the 'mainstream' economy, while in Newcastle, the emphasis is on incorporation into 'mainstream politics'. This reflects in part the geographical differences between the two areas. In North Tyneside, political life is being transformed into a model which parallels a particular type of economic mainstream, accustomed to a partnership between business and state support. In Newcastle, there are attempts at widening mainstream politics to include participation from the neighbourhoods. These different and limited outcomes from the community involvement in City Challenge are echoed in other studies of City Challenge experiences (National Council for Voluntary Organisations, 1993). But this takes place against a background of increasing alienation from 'mainstream' society. This arises in part because the dominant values of the 'mainstream' are not necessarily shared or useful to those having to survive on limited resources. Fundamentally, however, many in 'areas of concentrated disadvantage' recognise that the 'mainstream' economy and politics as currently constituted have little room for them. Thus unless the 'mainstream' economy and politics themselves transform, to incorporate an understanding of the real life circumstances of those living in areas of concentrated disadvantage, as discussed in the first part of this paper, incorporation of residents of such areas into the mainstream is a chimera. In this context, a positive outcome of both City Challenge programmes may be that business representatives understand the complexity and difficulty of the lives of neighbourhood residents.

Note

1. These accounts draw on our involvement in baseline monitoring work for the Newcastle City Challenge (Healey et al., 1992), evaluation and monitoring work for the North Tyneside City Challenge, survey of the emerging policy processes in Newcastle City Challenge undertaken in 1992 (Davoudi and Healey, 1994), work undertaken for Tyneside TEC (Davoudi, 1993), senior student projects in City Challenge (notably The Partnerships and Processes of City Challenge, 1992) and participant observation through membership of local boards.

Acknowledgements

We are grateful to those at Cardiff Seminar, to colleagues at Newcastle and to several of those involved in the two City Challenge initiatives for their comments on an earlier draft of this paper. A slightly different version of this paper is published in Environment and Planning C: Government and Policy, 1994, Vol. 12.

12

A COMPARATIVE ASSESSMENT OF GOVERNMENT APPROACHES TO PARTNERSHIP WITH THE LOCAL COMMUNITY

Annette Hastings and Andrew McArthur

Introduction

Community participation and partnership has become a dominant theme of urban regeneration in the 1990s. The direct involvement of the residential community in regeneration activity represents the latest strand of thinking in the search for more effective approaches to area regeneration. A less charitable interpretation, however, would be that involving local communities is merely the latest fashion in a lengthening series of government responses to urban deprivation. Does contemporary emphasis on community participation indicate that government is closer to an effective policy formula for urban regeneration? Or, are agencies involved because they feel they must be seen to be working in this way, and have little intention, or expectation, of changing their priorities or the way decisions get made?

Although community involvement is high on the agenda of urban policy in the 1990s, the idea that local people have an active role to play in urban renewal is certainly not new. In the 1960s the public was given new opportunities to participate in the process of town planning. Tenant control and ownership in the field of social rented housing has become widespread with the rapid growth of community-based housing associations and co-operatives throughout Britain (Clapham and Kintrea, 1992). Community credit unions are another example of local people taking responsibility for service provision. Although less well known than their co-operative counterparts in the housing field, there is a growing interest in their potential contribution in low income neighbourhoods (McArthur, McGregor and Stewart, 1993; NCC, 1994). A further form of community-owned and controlled enterprise is community business. Since their origins in the late 1970s in the West of Scotland, community businesses have spread throughout the country. In many regions local authorities have established specialist agencies to promote and fund their development (Pearce, 1993). The track record of community businesses has been less impressive than community-based housing associations and credit unions (McArthur, 1993).

Nevertheless, hundreds exist throughout Britain and provide valuable jobs and services under community-owned structures in low income neighbourhoods.

The list of innovative, entrepreneurial community activity extends well beyond these examples, and into the arts, health and so on (see Willmott, 1989). Perhaps then it should be no surprise that policy-makers are turning towards local communities to assist in area regeneration. However, much of existing community initiative is concerned with providing single services, some of this lasting only as long as government grant support continues. No doubt through many of these initiatives local people are making important contributions to the quality of life in their communities. However, this is not the type of community involvement we are dealing with in this chapter. Our interest is in situations where community representatives are working alongside government and other agencies to develop and oversee the implementation of an area regeneration strategy. It is in this respect that local communities have become partners in the regeneration process in what we have termed 'community partnerships'.

This chapter reports on research carried out in Scotland as part of a British-wide assessment of community partnership initiatives. In 1993 we found over twenty initiatives which could be described as community partnerships operating in Scottish towns and cities. Most of these initiatives began in the late 1980s and early 1990s, indicating that this is an approach which has gathered momentum in recent years. Contemporary urban regeneration is not only multi-agency but also multi-sectoral. Local authorities participate in all of Scotland's community partnership initiatives, in the great majority of cases working with the national housing agency, Scottish Homes, and Local Enterprise Companies (LECs). In Scotland, LECs, which have a wider remit than the Training and Enterprise Councils in England and Wales, are responsible for economic development, vocational training and environmental improvement. Health Boards are also involved in a number of initiatives, as is the private sector, usually in the form of a local Business Support Group (BSG). BSGs bring together a range of representatives drawn from local employers. Their function varies, but typically they provide assistance with vocational training courses and some have been involved in initiatives aimed at local schools. There were around a dozen BSGs operating in Scotland in 1994.

The complexity of community partnership initiatives poses a number of methodological issues for evaluators. The problems inherent in assessing any policy initiative which involves a number of agencies and actors with competing perspectives, interests and agendas are brought into sharp focus in partnerships. Multi-agency, multi-sectoral partnerships embody and give prominence to the process of negotiation and compromise inherent in the production of policy and strategy. There is therefore a need to use an evaluative method which recognises the competing and interest-bound agendas involved in any analysis of their outcomes. Pluralistic evaluation, as first articulated by Smith and Cantley (1985), is an account of the policy process which attempts to understand and analyse the potentially competing and subjective viewpoints of the key interest groups. Central to the conceptual framework on which pluralistic evaluation depends is the argument that success is a pluralistic rather than unitary measure. It should be analysed using the broadest range of data sources available and take account of the full range of interests with regard to the initiative.

The method adopted in the research on which this chapter reports attempts to grapple with the practical difficulties of constructing an evaluation which reflects multiple versions of reality. It employs many of the usual means of gathering qualitative data such as in depth semi-structured interviews, analysis of documentary material and observing meetings when access to these can be negotiated. However, particular attention has been paid in the evaluation of this information to the pluralistic character of the actual initiative. For example, efforts are made to ensure that written documentation is gathered from as wide a range of sources as possible and that the 'official' outputs of an initiative are not automatically accepted as an accurate version of events or outcomes. Interviews with consultees are conducted in such a way as to probe for the expectations which key stakeholders have of the process, particularly of the role which local communities play in the initiative. They also attempt to encourage individuals to indicate the value they attach to particular outcomes and to articulate how they judge 'success'. It should be noted that these are issues which people often find difficult to articulate, and as a means of identifying the underlying value systems and assumptions which inform how people perceive an initiative, it is not always a satisfactory method. However, a critical aspect of the research method used in this study, and which goes some way to addressing the problem, is the circulation of a full and confidential draft report to the consultees from whom comments are then invited. This generates both written and verbal feedback and a further round of interviews is arranged with consultees if required.

The second phase of consultation described above is designed to address a number of issues. It provides an opportunity to cross-check factual information, such as the chronology of events, which partner was responsible for an idea, the resourcing of the initiative, and so on. A consultation process of this sort also imposes a useful discipline on the researchers. It helps to ensure that the report focuses on arguments which can be substantiated from the information obtained during the original round of consultations, or from the documentation available. However, it may also mean that researchers will be overly cautious and avoid presenting an assessment which is critical of an initiative or of particular players.

There are two further benefits of the above element of the research method. First, it puts on display the range of perspectives and assessments offered by consultees for others to comment on and criticise. Allowing consultees to challenge the perspectives of other stakeholder groups proved a valuable way of establishing some of the underlying assumptions which inform interviewees' opinions, expressed in the first round of consultations. Some consultees articulated their own agendas and assumptions more explicitly than in the original interview as a conscious or unconscious consequence of arguing against other perspectives identified in the document. The final report is therefore enhanced as a result of the comments and critiques offered on the consultative draft, and the researcher obtains a more in-depth understanding of multiple perspectives involved in the initiative.

Second, it recognises that the research team inevitably becomes a stakeholder in the process of evaluation, and the relationship between the researchers and the evaluation is revealed to the other stakeholders in the partnership. Stakeholders can have differing views about the purpose and meaning of the evaluation and may not attach the same value to the findings (Rossi and Freeman, 1993). The assessment offered by the researchers may not please everyone involved. Thus, this second round of consultations gives consultees the opportunity to question and challenge the researchers' values and assumptions. In this way, another dimension is added to the pluralistic framework of the evaluation in that the perspectives of the researchers are exposed to scrutiny.

The remainder of this chapter focuses on an assessment of two distinct models of community partnership which are currently underway in Scotland. The first is local authority-led and involves a devolved and decentralised local government approach as represented by the joint social and economic initiatives in Drumchapel and Easterhouse, two of Glasgow's peripheral housing estates. The second is led by central government and comprises the Scottish Office's Urban Partnership Initiatives which are taking place in four of the country's problem outer estates: Wester Hailes in Edinburgh; Castlemilk in Glasgow; Ferguslie Park in Paisley; and Whitfield in Dundee. The experience of one from each of these models (Drumchapel and Wester Hailes) will be explored. However, before doing so, some of the arguments which surround the issue of community partnership are considered, particularly since these illuminate some of the conflicting perspectives about what the intended rationale for community partnership is.

The Case for Community Partnership?

A number of arguments can be found in support of partnership approaches and closer working links between government and local communities (see, for example, Donnison, 1991; McKintosh, 1992; Bailey, 1994; and McGregor *et al.*, 1992). With specific reference to involving local communities in the area regeneration process, some of the main arguments are:

• When local communities are given a sense of ownership over the regeneration process they are more likely to 'guard' or demonstrate stewardship over the improvements made. In this way regeneration becomes more sustainable and cost effective.
• 'Better' and more effective regeneration strategies result from an input by people who are experiencing the problems at first hand. Their voice is important in getting the story right regarding local needs and priorities.
• Synergy, or added creativity, is generated by bringing several partners, including the community, together who share different outlooks and perspectives.
• Community development and the empowerment of residents creates new organisational capacity on the ground. This adds to the potential for further local initiative and for productive links between local communities and external agencies to continue and outlast time-limited regeneration initiatives.

The problem with many of these arguments is that they tend to be speculative and untested. They are counter-balanced by a number of concerns:

• In the context of a growing *mixed economy of welfare* (Association of Metropolitan Authorities, 1993), with new market mechanisms, the extension of voluntary and private service provision and cutbacks in statutory provision, is the growth of community involvement merely asking poor communities which are already carrying a severe economic and social burden to do more to solve their own problems?

- Many community groups perform a worthwhile campaigning role, maintaining a critical perspective on the role of government and service providers. Is partnership a means through which, deliberate or otherwise, valuable protest will be undermined as a result of the incorporation of dissenters into structures and cultures which dampen their critical perspective?

- The above point raises the issue of motivation behind partnership initiatives. The vast majority of existing community partnerships represent top-down initiatives imposed by external agencies. It is hard to find examples which are genuinely led by residents from the grassroots. Are partnership initiatives a commitment to a new community-based and participative style of working, or are they a way of manipulating the views and demands of local communities?

- There are also practical questions about how best to approach a partnership initiative. External agencies frequently have limited local intelligence about local community networks. The local situation is often complex, with many community organisations, each with its own agenda, in competition with one another for influence and resources. Who should represent the community in the partnership? Usually this is a decision which is made quickly, given the pressure to make progress on the ground. There is also a related issue of supporting and resourcing the community's involvement in the partnership. Will additional resources help community representatives make an effective contribution and avoid the obvious danger that the partnership might be an imbalanced and unequal one? If so, what is required?

- Apart from these practical issues about engaging, organising and resourcing the community, what difference will community involvement actually make? What does community partnership bring to the regeneration process? The answer may not be straightforward. These two questions can actually deal with different issues. It is possible that involving local communities in decision-making processes will lead to different, or modified, policies than would have been the case had the community not been a partner. However, it does not necessarily follow that the other partners will agree that the influence on regeneration policy exerted by the community is best for the area and leads to more effective regeneration. Here we have a seed of potential conflict stemming from the different interpretations of 'success' which the various participants might bring to the partnership table.

Before proceeding with the case studies it is necessary to be explicit about the criteria which are being used in this research project to evaluate the success or effectiveness of community partnership initiatives. Our interest goes beyond explaining the varying viewpoints about the successes and shortcomings of individual projects. The research is premised on the assumption that for community partnership to be meaningful it should be possible to trace the difference which community involvement makes to the partnership process and to the outcomes of the process. Our view is that, in the absence of an observable effect, we need to be cautious about accepting that a partnership initiative involves the community as a genuine partner. This is not to oversimplify the process of policy development or delivery. In partnerships in particular it is notoriously difficult to trace the relationship between the inputs of any one partner and the outputs of the partnership as a whole. We do not necessarily expect to extricate a list of achievements or a catalogue of battles won by the community, but if community involvement has brought anything additional to the partnership initiative then we would expect to find some evidence of this.

The Drumchapel Initiative, Glasgow

We begin our account with Drumchapel, an estate with a population of around 19,000 on the north-western edge of Glasgow. Drumchapel is the smallest of four large peripheral estates built by the local authority to rehouse people displaced by comprehensive redevelopment programmes in the inner city during the 1950s and 1960s. The estate's fortunes have declined dramatically since it was built. It has suffered heavy population loss and traditional sources of employment, such as the former shipbuilding and heavy engineering industries in the nearby town of Clydebank, which lies only a mile from the south-western edge of the estate, have contracted or collapsed. In January 1994 registered unemployment stood at 30 per cent (Strathclyde Regional Council, 1994). Unemployment is the issue which concerns most people in Drumchapel (rather than drugs or crime as in some other areas). Poor health and poverty are related problems. It was reported on the BBC national news during January 1994 that Drumchapel residents can expect to live on average ten years less than their neighbours in the affluent Glasgow suburb of Bearsden which borders the estate.

The Drumchapel Initiative was announced in 1985 as a three-way partnership between both tiers of local government and the residents of the estate. The partnership is a statutory sub-committee (or joint committee) of the Policy and Resources Committees of Strathclyde Regional Council and the City of Glasgow District Council. This body, the Area Management Group (AMG) is formed by three councillors from each authority and three local residents. All nine have equal voting rights.

In addition to the AMG and its various sub-groups, two other elements are important to the Initiative's structure: the Community Organisations Council (COC) and Drumchapel Opportunities (DO). These were not in place at the start, but have grown out of the core Initiative and represent its community development and economic development arms respectively (Figure 12.1).

The partnership set out to improve every aspect of life in Drumchapel. This *comprehensive* orientation is consistent with other partnership initiatives in Scotland. In Drumchapel, however, the partnership proceeded on the basis of a fairly limited range of partners. As demonstrated below, the more recent initiative in Wester Hailes draws together other important players in the urban regeneration field. The AMG was charged with three main functions. First, the control of a community budget and a budget for environmental spend which, combined, recently totalled £0.75 million per annum. Second, the consideration of Urban Programme applications. Third, and most important, it was intended to act as a powerful advocate for the area by using local knowledge to exert a strong influence on the key departments of both local authorities, although it

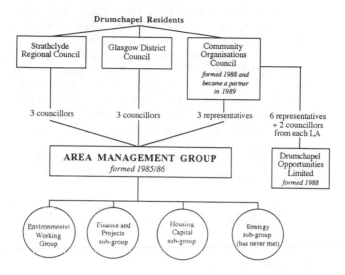

Figure 12.1 Organisation structure of the Drumchapel Initiative.

should be stressed that, despite a degree of devolved power, the AMG cannot instruct local authority departments to alter their broad policies to suit the particular needs of Drumchapel. The consultees disagreed over the extent to which the AMG has been able to influence local authority departments. One view was that departments merely reported their plans and activities to the AMG for rubber-stamping. The contrasting view was that a process of negotiation between the AMG and departments was possible. There is an obvious relationship between the limitations of the AMG with respect to the local authorities and the role which community representatives on the AMG could have on what happens in Drumchapel. This will be explored further later in the chapter.

Engaging and working with the local community

Drumchapel has a reputation within the city as a resilient community with a fairly high level of community activism. There are currently around 150 local groups on the estate ranging from tenants' groups, self-help groups and community-based housing associations and co-operatives. Prior to the Initiative there was no umbrella community structure for these organisations. Although the authorities began the process of engaging the local community in the new Initiative in a fairly *ad hoc* and informal way by calling together a group of residents and finding six local people willing to work with officials and councillors in developing plans for the Initiative, the structure for community involvement became more formalised shortly afterwards. Nine of the pre-existing groups (including the two Community Councils) were brought together as a Community Organisations Committee which held a public meeting formally to elect three representatives on to the AMG. By 1988, this group had evolved into the Community Organisations Council (COC) which has around

73 of the 150 or so local groups operating in the estate affiliated to it. In 1989 the COC became a formal partner of the Initiative and has been responsible for organising the election of the three community representatives who are then nominated to the AMG. An open election takes place every two years, with any Drumchapel resident aged over 16 years being eligible to vote.[1]

The extent to which local groups are able to maximise their input into the AMG would seem to be dependent on two things: the depth of funding enjoyed by the COC and the way it organises its participation in the Initiative. The COC's support for community participation in the Initiative is delivered through the Community Resource Unit (CRU) which had six staff and a budget of £123,000 per annum in 1993, funded through the Urban Programme. The COC has also developed a variety of roles, including providing services on the estate, which are discussed later.

The CRU was initially set up as a way of supporting the community in identifying issues, articulating demands and influencing the agenda of the Initiative. The COC has attempted to do this by linking the community representatives on the AMG closely to the elected board of the COC. The board meets with the community representatives prior to AMG meetings to discuss the agenda and, if necessary, to mandate the community representatives to put forward an agreed point of view or vote in a particular way. In effect, a community 'block vote' is created. The Managing Director of the COC describes this as a form of double accountability. Community representatives are accountable to the whole Drumchapel community through the electoral process. However, they are also accountable through the board of the COC to its membership and this way again to the local community.

To assess how the community's involvement in the Drumchapel Initiative has worked in practice, we need to consider how the relationships, processes and structures which characterise the Initiative have facilitated residents to influence decisions taken about Drumchapel by the local authorities. If we consider the original engagement of the local community, it would seem that the way this was organised influenced the potential role of the community in determining the structures and priorities of the Initiative.

The CRU was not established until 1987, two years after the Initiative had started. During these two years there were no dedicated resources available to support the community's involvement. It was taken for granted that the existing base of local activism would be sufficient to allow the community to influence the early stages of the Initiative. Residents were expected swiftly to come to terms with the planning of a major joint local authority initiative. Although the representatives have positive recollections of their early experience, they were unable to influence how the Initiative developed and there is little sign that they played a role in setting priorities. Community representatives were hard pressed during the first couple of years. One member recalls attending over 300 meetings in one year. These pressures have eased as the constituency of activists has increased and resources have gradually been put in place to support the community's involvement.

Early problems in working relationships between local residents and the councillors involved emerged. AMG meetings are held in public and agendas are distributed to local organisations. For the first couple of years, observers from the floor were allowed to ask questions. This led to lengthy sessions and difficulties in dealing with agendas. Sometimes councillors would be faced with awkward questions. As a result of councillors' difficulties with this open approach, the legal standing of the AMG as a sub-committee of both local authorities was invoked. Strict adherence to normal procedures and standing orders meant that questions could no longer be put from the floor, but had to be submitted in writing prior to the meeting. The result was to raise unease amongst local people that opportunity for debate which they had expected the AMG to provide was being restricted. At one point, the community representatives withdrew from the AMG in response to what they saw as an autocratic style among councillors.

Many of these early difficulties passed and, in general, relationships have been characterised by a lack of confrontation and conflict. Only occasionally do decisions go to a vote. Indeed, where more serious disagreement has emerged, it has tended not to be between the community representatives and the councillors, but between the District and Regional councillors themselves. For the community representatives these disputes, more than the formal structure of the Initiative, have provided opportunities to exert an influence. For example, a Regional Council proposal for a road through a local park was opposed by the District councillors. The community representatives voted with the District members and successfully blocked the plan. The scope for the community to exert such leverage has been reduced in the latter stages of the Initiative. The tension apparent in the early days between the two sets of council members has been reduced by an unwritten and almost tacit agreement that Regional councillors will not comment in detail on District services, and vice versa.

Over the seven years the Initiative has been in progress there have been few instances of the AMG challenging decisions made at the centre. Basically, the political system has carried on as usual, with councillors lobbying the centre informally on behalf of their constituents' interests in much the same way as would be expected in the absence of the partnership initiative. Community representatives in particular argue that the AMG rubber-stamps, rather than challenges, decisions already taken by central departments.

Impact of community participation in the Drumchapel Initiative

The lack of conflict between community representatives and councillors on the AMG in the more mature phase of the Initiative has already been raised. This appears to be partly a consequence of the limited expectations which participants now have of the scope of AMG to influence the policies of the two local authorities. In general, important policy decisions have not been on the table for discussion. Furthermore, there is little evidence to indicate that the community representatives and the main community organisation (the COC) have sought to challenge this situation.

An illustrative case concerns the closure of two of Drumchapel's three secondary schools, including the only Roman Catholic secondary in the area. The Regional Council's plans were widely unpopular among residents and District members. Staff and activists of the COC helped mobilise local opinion and during the ensuing campaign 1,300 people wrote to the Education Department protesting against the school rationalisation proposals. Despite active lobbying behind the scenes from councillors, the AMG did not as a body debate the issue or press for an alternative course of action. District members were reluctant to step into the 'territory' of Regional members, and Regional members were unwilling to break with the party line. Of particular importance was the fact that community representatives, despite the position the COC had taken, did not push for a vote on the issue. Arguably, they could have tried to use the District members' opposition to have the AMG formally oppose the plan.

Councillors are able to explain why, from their point of view, the closing of the schools in Drumchapel was not an issue with which to test the muscle of the AMG. The school rationalisation decisions were rooted in strategic, region-wide policy and based on the belief that pupils in schools with small rolls were in danger of being deprived of a proper education. Open confrontation between the AMG and the regional council on this issue could well have exposed the local participants as ineffective and undermined the credibility of the Initiative as an advocate for Drumchapel.

The reasons why the community representatives did not force the school closure issue at the AMG are less easy to explain, especially given the depth of local feeling against the proposals and the fact that the COC had helped to mobilise the local campaign. Indeed, in the draft report we circulated to consultees as part of the research we suggested that the community representatives had missed an opportunity to use the AMG as a forum to contest the regional council's proposals. However, this was one example of where the second round in the consultation process resulted in people contesting the researchers' tentative conclusions. Staff of the COC and some community activists stressed that they did not interpret the experience in this way, although the only explanation offered was 'that the time was not right to test out whether the AMG could have an influence'.

Notwithstanding the impotence of the AMG with regard to the school closure issue, there is one notable occasion when the AMG has successfully challenged the regional council's strategic planning policy. A large and strategically important industrial site on the edge of the estate had lain vacant since the closure of a Goodyear tyre factory in 1980. The regional council's structure plan had identified the site for industrial use. However, following lengthy negotiation with the AMG, the COC and Drumchapel Opportunities, the regional council revised its structure plan and granted planning permission for a mixed-use development based around a retail superstore. The united front presented by the Initiative, coupled with behind the scenes lobbying by the regional members of the AMG, is credited with bringing forward the development of the site. The influence of community participation on this change, however, is debatable. Although community representatives were involved to an extent, officials and councillors played a major role in persuading the regional council to alter its strategic plan.

There are few other examples of where the Initiative, through strong advocacy, has altered strategic decisions, but other impacts can be identified. Regional council departments have adopted a more local focus with the appointment of area officers for education and roads. Social Work also appointed an assistant district manager to the Initiative. At the level of the district council, officers in planning and parks and recreation, little known in the area prior to the Initiative, have become more accessible to local groups and community representatives. Again, however, we must be cautious about attributing these changes entirely to community involvement as both Strathclyde Region and Glasgow District are councils with a long track record in experimenting with decentralisation measures.

When we turn to working procedures, a number of adjustments are attributable to the participation of community representatives. Reports, and the language of professionals, have become less jargonistic. Many of the people stressed this was the outcome of officials working more closely with resident activists over a number of years. Furthermore, when preparing reports, officials from central departments now sometimes approach community representatives directly for their views.

So far, the discussion has concentrated on the AMG. However, this misses out additional activities of the community development arm (the COC) and the role of the economic development arm (Drumchapel Opportunities) of the Initiative. Although set up as a mechanism to engage and support the community representatives, the COC has become a direct provider of numerous local services. It is involved in management and accountancy training for local organisations, community care services, welfare rights, food co-operatives, furniture recycling, and an architecture service, and runs a local theatre and community facility. In 1993 the organisation's turnover and assets were estimated to be in the region of £1.5 million and £1.3 million respectively. The COC employs around 60 people, around two-thirds of whom live locally. The board of fourteen directors are all local people representing community organisations or are individual members of the COC.

The second of these organisations, Drumchapel Opportunities, assists hundreds of local people each year through training schemes and job placement work. It is widely regarded as a model local training and employment initiative and similar approaches have since been followed in other regeneration areas. The organisation is also a significant local employer in its own right, with 65 staff, well over half being local residents. Half of the board are residents nominated by the COC; a higher proportion of local people than in other similar organisations elsewhere in the city.

There are a number of issues raised by the expansion of the community's role in the initiative beyond representation on the AMG, to participation in, and possible control over, organisations providing local services and employment opportunities for local people. An assessment of the community's role in bodies like the COC and Drumchapel Opportunities has to grapple with difficult issues. Are the services provided additional? Are they more sensitive to local needs as a result of the involvement of local people in their management and delivery? Does the involvement of residents in community-based organisations of this sort develop new skills and confidence among activists? Does the experience and confidence developed encourage residents to become involved in new activities and initiatives? And, crucially, what is the local community's real influence over fairly large and complex organisations which employ teams of professional staff, even when locals are well represented within the management structure of the organisation? These are some of the questions which could inform an ongoing research agenda in the increasingly fashionable area of community partnership.

The Wester Hailes Partnership, Edinburgh

The housing estate of Wester Hailes lies on the western edge of Edinburgh and looks out towards the Pentland hills. It is home to around 11,000 people housed in multi-storey and walk-up blocks of flats built in the 1970s. The estate was designed around cars rather than people, with no pavements alongside roadways but with extensive walkways and bridges for pedestrians and car parks instead of gardens or landscaped space. The estate has its share of social and economic problems. In 1991 unemployment stood at 22 per cent compared to a city average of 8.6 per cent (Wester Hailes Partnership, 1993). In 1989, around 60 per cent of local households were living on incomes of under £5,000 per annum (MC Economics Limited, 1991) placing them in the bottom 20 per cent of UK household incomes. Whilst it has a reputation for crime and drug abuse in Edinburgh, it is not the most disadvantaged area of the city.

In 1988 the estate became the focus of one of four flagship Urban Partnership Initiatives announced by the Scottish Office in *New Life for Urban Scotland* (Scottish Office, 1988). The Wester Hailes Partnership is intended as a demonstration project designed to explore the potential for a new co-ordinated and comprehensive approach to regeneration within the existing legislative framework. A distinguishing feature is that the local community is a full partner in the process. There are nine partner organisations in all (see Figure 12.2) representing the public, voluntary and private sectors.

The Partnership is structured around a board comprising community representatives, local councillors, and senior officials representing the other partners. The board is chaired by a senior civil servant from the Scottish Office and meets every two months. In addition, there are five sub-groups focused on particular strategic issues. A business support group and an urban programme panel complete the picture. A partnership team of fifteen staff is employed to implement the strategy and monitor its progress. The Partnership has no control over a devolved budget. It is expected that the partners, who participate voluntarily, will spend according to the agreed strategic aims of the Partnership. Community participation is structured through the Wester Hailes Representative Council (Rep. Council) which was set up in 1981, some seven years before the Partnership came to the estate. At the outset of the Partnership, Wester Hailes had a strong and well-resourced community organisation firmly established with a history of protest and campaigning. This contrasts with the situation in Drumchapel where the creation of the umbrella community organisation (the COC) took place after the Initiative had been established. This would appear to have substantially affected the influence the Wester Hailes community has been able to exert on the Initiative. The Rep. Council is funded through the Urban Programme and it is the best resourced community structure in Scotland, employing nineteen staff. Links between the Partnership and the Rep. Council are maintained through twelve spokespersons who each have a specific policy brief. The spokespersons attend Partnership sub-groups relevant to their brief and are responsible for talking to their brief at Partnership board meetings. In this way, individuals build up expertise in a particular field and there are clear lines of responsibility within the Rep. Council for particular issues.

Although the Partnership has been in operation for five years, there is a lack of consensus among participants about what its purpose is. The main differences are between those who represent the Wester Hailes community and those who represent the other partners. The other partners perceive the Partnership as being largely about service delivery issues: the co-ordination of services, avoiding duplication and tackling problems simultaneously. The involvement of the community is a means by which the delivery of services can be improved and adjusted to meet local needs.

Representatives of the Wester Hailes community, however, see the Partnership in different terms. They place more importance on the quality and match of services to local needs than on inter-agency co-ordination. For them the Partnership has (or should have) a sense of collective responsibility and a remit to 'run' the regeneration process. They perceive their own role as a chance for local people to become major players in shaping what happens

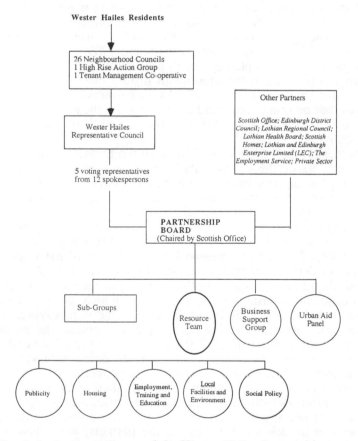

Figure 12.2 Organisation structure of the Wester Hailes Partnership.

to the estate. These two contrasting perspectives make the evaluation process difficult. The Scottish Office is clear about the scope and scale of the Partnership, although there is an acknowledgement that the publicity which has surrounded the Partnerships may have raised expectations. However, since the research is concerned with charting the extent to which these initiatives have facilitated local people to influence local decision-making, the focus of the discussion is therefore on strategies adopted by the Rep. Council to shape the Partnership.

Engaging and working with the community

At the outset of the Partnership, the Rep. Council was approached by the Scottish Office to represent the community. There was recognition that without its support it would have been very difficult for the Partnership to proceed in Wester Hailes: a political reality which the Rep. Council recognised gave it bargaining power. It then set about to try and establish a set of rules for the Partnership and define its purpose by drawing up a document which set out the terms under which the Rep. Council would be involved. The general thrust of the document is captured in the following words: 'The community will do, not be done to – processes and policy determination must be led by the community. The agendas of other members of the Partnership are subordinate to this.' (Wester Hailes Representative Council, 1989, p. 5.)

The strong position adopted by the Rep. Council needs to be borne in mind when trying to understand the subsequent history of the relationship between the Rep. Council and the other partners. As the Partnerships were never intended to be led by the community, the kind of involvement the Rep. Council is seeking is not the perspective the other partners have of the role of the local community in the Initiative.

The Partnership board is a central focus of the Partnership and oversees and monitors progress on the strategy. Its bi-monthly meetings represent the only time that representatives from all the partners come together. Opinions are divided, however, over what should be achieved at Partnership board meetings. Community representatives are critical of the fact that the Board has developed a role of merely ratifying, or rubber-stamping, decisions made by the sub-groups. Other partners, however, consider this to be an entirely appropriate role for the board. The Rep. Council has pressed formally for board meetings to be used as a forum for debate and discussion, and in practice the spokespersons have tried to use Board meetings to call other partners to account, to extract funding promises or to force partners to keep promises which the Rep. Council feels they have made. In using the board as it does, the Rep. Council is engaged in a process of continually trying not only to renegotiate the purpose of the Partnership board, but also to shift the orientation of the Partnership towards a position where the community is 'running' the decision-making process by exerting a strong influence over its constituent partners.

The Wester Hailes experience suggests that there may be a level missing from the organisational structure of the Scottish Office model of the Partnership, perhaps another tier at which the strategic business under way throughout the sub-groups can be debated between all the partners, and where the actions of individual partners can be queried or challenged. Until both of these requirements are met, it is likely that the Partnership board in Wester Hailes will continue to be an arena in which frustrations surface.

Conflict over the purpose of the Partnership board also reflects a lack of clarity over what community representatives are expected to contribute and at what stage in the process. It also suggests a fundamental difference between the community and the other partners regarding the overall purpose of the Partnership. A formal committee, geared principally to overseeing and ratifying decisions, is more appropriate to a model which emphasises the co-ordination of services and the smooth delivery of policies and programmes. However, if the purpose of the Partnership is, as the community organisation sees it, to *run* the regeneration process, then there is a need for a component to the structure which gives the community representatives the teeth necessary to make demands of the partner agencies.

Impact of community participation through the Partnership

The influence of community involvement on the Partnership has, so far, been greatest in relation to housing issues, particularly the tenure structure of the estate. The Rep. Council's *Community Housing Strategy* (Wester Hailes Representative Council, 1989), drawn up in the first year of the Partnership, argues that the housing needs of existing residents will not be met by the large-scale introduction of housing for sale. A resistance to the expansion of owner occupation has meant that 85 per cent of the housing stock remained in the ownership of the local authority in 1993 (Wester Hailes Partnership, 1993). This situation contrasts markedly with the pattern of tenure change which has taken place in other Partnership areas. For example, in Castlemilk in Glasgow, the proportion of the housing stock owned by Glasgow District Council stands at 69 per cent, with 27 per cent owned by community-based housing associations or co-operatives, and 4 per cent owned privately (Castlemilk Partnership, 1994).

The Rep. Council considers it has also resisted government's intentions to use Wester Hailes as an estate within which to experiment with high rise accommodation through changes to management and allocations systems in order to show other local authorities that high rise could be made to work. A reduction in high rise homes became one of the objectives of the strategy document, *Realising the Potential*, and demonstrates, in the eyes of community activists at least, that community involvement was able to make an early contribution to the shape and direction of the housing strategy, and alter the direction of the government's agenda.

The involvement of the Rep. Council has widened the agenda of the Partnership beyond its initial focus on physical and economic issues. Although the original document announcing the four Scottish Office Partnerships envisages a comprehensive approach, the Partnership strategy in the Wester Hailes, *Realising the Potential*, did not consider social concerns. However, in the autumn of 1990, the Rep. Council undertook a consultation exercise to identify what the social priorities should be. This highlighted four areas of concern: education, health, leisure and recreation, and social welfare. In March 1991 the Partnership established a social policy sub-group to address these issues. Since its inception the scope of the social policy sub-group has changed.

Some concerns have been reallocated to other sub-groups: education to the employment, training and education sub-group and leisure and recreation to a community facilities and environmental sub-group. Other social concerns identified as the social policy sub-group developed, such as crime and childcare, have also been integrated into the mainstream operations of the Partnership. Hence, issues initially marginal to the Partnership's original strategic agenda, but of concern to the local community, have been integrated into the mainstream business of the Partnership.

It should be noted that community participation may produce some unwelcome policies from the point of view of other participants. The issue of tenure diversification stands out. The Rep. Council's opposition to introducing significant amounts of housing for sale is in direct contradiction to one of the key objectives of one of the funding partners (Scottish Homes), and is opposed to central government policy. There is the potential for significant tension to develop around this issue and it remains to be seen how this is played out in the remaining years of the Partnership.

The Wester Hailes Rep. Council membership is composed of twenty-eight geographically based representative organisations, twenty-six of them neighbourhood councils. Each neighbourhood council represents a small part of the estate comprising around 200 households. The structure has been in place since 1992 and replaces the old umbrella structure which brought together all sorts of organisations in the estate: a mixture of representative, issue based and service delivery organisations.

Perhaps an unexpected area of impact concerns the consequences participation has had for the established structure of community organisation, namely the Rep. Council. In Wester Hailes the existing Rep. Council appeared an ideal structure to link up with. However, strong dissatisfaction soon developed over the ability of the organisation to represent grassroots interests. There are signs that all was not well prior to the Partnership, but the Partnership process brought concerns to a head and precipitated the reorganisation of the Rep. Council.

The result of the reorganisation is that a number of new people have become active in the Wester Hailes community. The involvement of these new activists appears to have allowed the organisation to retain its critical edge and to have avoided the dangers of incorporation. However, it also meant that some of the established activists were sidelined, stimulating the Rep. Council to look at how these more experienced people can continue to be involved and to ensure that their expertise is not lost.

There are a number of important factors which have influenced the character and impact of resident participation in the Wester Hailes Partnership. The Rep. Council was a mature and established community organisation prior to the Partnership and was able quickly to take a position on the Partnership and to participate with a strong and independent voice in the development of its strategy. In the early days it influenced key decisions such as the demolition of high rise flats, and over time, has been able to introduce a more social emphasis. The Rep. Council has also retained its original purpose as an advocate for the estate and has avoided the dangers of incorporation. However, by adopting a fairly critical stance within the Partnership, it has sacrificed a more comfortable relationship with the other partners and has not fully accepted the consensus-orientated model of partnership which the Scottish Office wishes to promote.

A final point concerns resources. The Wester Hailes Rep. Council is secure and well funded. Community participation is more generously resourced here than in many other regeneration areas. In Wester Hailes the community has something like the backup the other partners can call on in the form of professional staff and civil servants. Nevertheless, the community still feels that it is unable to make the scale of contribution it would wish. In 1994 the Rep. Council was considering whether it would be more effective to focus on only a few of the Partnership's areas of strategic interest. Given that such a well-resourced community organisation has these concerns, there may be grounds for a general rethink of the scale of resources appropriate to support community organisations participating in regeneration initiatives, or of the range of issues around which this participation takes place.

Conclusions

The research reported in this chapter represents work conducted in the early stages of a large nationwide study into the nature and effectiveness of community partnerships. The discussion has emphasised that the complexity of community partnerships, involving partners and participants with varying objectives and different perspectives, makes an assessment of them more difficult. Focusing on how partnerships facilitate local resident influence over the decisions taken about their locality, means that the researcher is concentrating on one particular rationale for community participation which not all participants share. Other partners, for example, perceive the process as primarily a consultative one, rather than one based on a commitment to shared decision-making with residents.

The research method has tried to accommodate differences in perspectives across the partners by circulating a report outlining the researchers' preliminary assessment of the Initiative and inviting further comment and feedback from those consulted during the study. This dissemination process has confirmed the potential for this interventionist research method to create or amplify tension or conflict, simply by displaying the different perspectives on the Initiative (see Means and Smith, 1988). The research method has also come under scrutiny from consultees, and some have criticised the approach and output as being overly dependent on qualitative or 'anecdotal' information. It is clear from the feedback obtained that some of the consultees interviewed, from both the community and agencies, do not perceive 'research' to be qualitative and would be much more comfortable with quantitative information generating clear-cut conclusions.

In this chapter we have examined two different examples of community partnership. Drawing lessons to apply in other localities from these experiences is difficult. The Drumchapel model's potential for wider replication is limited. The reorganisation of local government in Scotland to create single-tier authorities removes the scope for this form of joint social and economic initiative taking place. Furthermore, the provisions of the Local Government and Housing Act 1989 rule out local authority committees extending voting rights to co-opted members and insist on a political balance on all subcommittees which reflects the political balance of the full council. However, the statutory instrument necessary to put these clauses into effect has not yet been brought forward by the Secretary of State for Scotland. In Wester Hailes, the Partnership is expected to carry on towards the end of this century. The discussion contained in this chapter, therefore, deals only with the first half of its life. The experience of community involvement may be quite different in future. Furthermore, certain assumptions about the value of community partnership, for example greater stewardship being shown by residents towards the local environment which they have had a hand in shaping, cannot be fully assessed over relatively short time-frames.

Despite these caveats, there are some important points which emerge from the two case studies which should be highlighted. The process of partnership is unlikely to be a smooth one. As Wester Hailes demonstrates, particularly where an established community organisation exists which is capable of voicing its own agenda, considerable tension can develop between the community and other partners. Perhaps this is no bad thing and we should look on situations where partners are faced with challenges and the need to reassess their priorities as evidence that the partnership is alive and genuine. Equally, in situations where there is little evidence of tension, and where the partnership is characterised by consensus and agreement, one should be cautious about interpreting this as success. The Drumchapel case study highlights how the partners can effectively reduce the scope for debate and advocacy by agreeing behind the scenes not to interfere with each other's strategic policy decisions.

Other observations concern structures for community participation. In Drumchapel a new community structure was set up to facilitate local involvement which, in turn, and unexpectedly, developed an important role in the life of the estate as a provider of numerous local services. In Wester Hailes this had not happened, but the umbrella community body which represents the area within the Partnership has, as a result of participation, had to contend with new pressures and demands emanating from the grassroots. The restructuring of local community representation, building from a newly established network of small community or neighbourhood councils, was precipitated by the existence of the Partnership. In Wester Hailes the community has managed to adjust and reorganise without great difficulty, but in other areas the impact of partnership for the community organisations involved could be more painful.

The case studies have indicated that community representatives have been able to influence different levels of decision-making ranging from the strategic (e.g. housing and social policy in Wester Hailes) to the administrative (e.g. informal links between officers and residents in Drumchapel). However, the potential contribution of community partnership to urban regeneration remains an open question. Furthermore, looking ahead, new questions will come to the fore. What happens to the community structures involved beyond the life of time-limited partnership initiatives? Will the capacity created for local people to participate directly in decision-making continue? Have new and productive relationships between local communities and external agencies been created that will be sustainable? It seems likely that researchers will be studying and writing about community partnership for several years to come.

Acknowledgement

The authors are grateful to the Joseph Rowntree Foundation for supporting the work on community participation and partnership upon which this chapter is based.

Note

1. In the first election to the Community Organisations Council in 1989, almost 600 people voted. Turnout doubled to 1,200 in 1991, but fell back to 800 in 1993. The 1993 vote represents about 10 per cent of the eligible electorate: not an insignificant turnout, but lower than that for the local council elections which averages between one-third and one-quarter of the electorate.

PART FOUR

The Role of Cross-National Comparisons

13

REVIVING THE VALLEYS? URBAN RENEWAL AND GOVERNANCE STRUCTURES IN WALES

Kevin Morgan

Introduction

At a time when capital is becoming ever more *global* in scope and character the growing interest in *local* and *regional* governance structures might seem perverse. On reflection, however, there is nothing perverse about this because while the causes of uneven economic development are global in origin, many of the effects have to be addressed at these sub-central levels. Recognising this is one thing, having the capacity to do so is another matter. This raises the problem of sub-central governance capacity, a problem which has reached crisis proportions in the UK, where the territorial division of political power is now so manifestly skewed towards the central state that little if any autonomous action is possible without the blessing of Whitehall.

This is quite clearly the case with local government, an institution whose powers have been severely handicapped by a combination of government legislation on the one hand and by the growth of unelected quangos on the other. With the exception of Wales and Scotland – where a modest degree of institutional devolution has been achieved – the UK lacks the regional institutions which play such a key role in orchestrating economic renewal in other EU countries. This regional institutional deficit needs to be redressed for both internal and external reasons: internally, to provide a strategic context for local economic development strategies and, externally, to enable UK regions to forge stronger links in Europe, where they are in danger of losing out to their better equipped EU counterparts (Audit Commission, 1991a).

The evidence from Europe suggests that the design and delivery of effective local economic development policies depend in no small way on robust local and regional governance structures, i.e. the nexus of sub-central institutional networks through which public and private resources are mobilised and deployed (Anderson, 1992; Cooke and Morgan, 1993; OECD, 1993). The question of local and regional governance structures looks set to become a highly contentious issue in the UK in the 1990s, not least because so many public service functions – in education, training, health, housing, urban renewal, enterprise support and so forth – are now being managed by bodies which are neither elected by nor accountable to the communities in which they operate (Stewart, 1993). This issue of democratic accountability needs to be considered as part of the urban and regional development agenda because, as I shall argue later, unelected and unaccountable bodies are unlikely to have the confidence of their local communities, and this saddles them with a major credibility problem.

The chief aim of this chapter is to explore the twin issues of urban renewal and governance structures in Wales. The focus on Wales is instructive for two reasons. First, the Valleys Programme, which aims to regenerate the traditional industrial heartland of the Welsh economy, claims to be one of the most ambitious urban renewal projects ever undertaken in the UK. Second, it is often said that the regional state apparatus in Wales provides an institutional context in which the problems of urban development can be addressed in a more strategic manner than is possible in the English regions; but to what extent, if at all, is this true? Before examining these issues we need to take stock of the changes which have been wrought in the spatial policy arena in recent years.

Territorial Policy in the UK: The Neo-Liberal Repertoire

Few parts of economy and society escaped the Thatcherite drive to create an 'enterprise culture' in the UK in the 1980s. Driven by a deep ideological aversion to the public sector, which was seen as inimical to 'enterprise', the Conservative governments set a high premium on rolling back the public sector, either by privatising public sector assets or, where this was not possible, by exposing them to the rigours of market pressures, e.g. compulsory competitive tendering in local government, market testing in central government and the quasi-markets in the education and health services. This pro-market ethos, which is being carried into the 1990s by the Major government, also informs the restructuring of territorial policy, which embraces urban, regional and local government policies.

Historically, *regional* policy has played the key role in the government's territorial policy repertoire. For successive governments since the 1930s regional policy was the chief instrument for maintaining inter-regional balance in the UK economy, and was therefore the litmus test for a government's commitment to 'One Nation' politics. Although the political commitment to regional policy began to wane in the late 1970s, it was not until the Thatcher era that this policy was seriously devalued. As we can see from Table 13.1, government expenditure on regional policy declined by 72 per cent in the decade to 1991/92, a clear signal that the government was becoming less *directly* involved in maintaining inter-regional balance.

Table 13.1 Regional policy expenditure (1992/93 prices)

	1981–82 (£m)	1991–92 (£m)	% change
England	818.0	157.6	(-80.7)
Wales	367.5	160.5	(-56.3)
Scotland	432.6	134.4	(-68.9)
Great Britain	1,618.1	452.5	(-72.0)

Source: Parliamentary answer (4 March 1993).

Although these expenditure cuts grabbed the political headlines, an equally important, if less perceptible, change was the new emphasis placed on 'natural adjustment' as a means of correcting inter-regional imbalances. In the Conservative scenario inter-regional imbalances in unemployment are deemed to be caused by labour market rigidities like national wage agreements. Hence the solution, in this view, is to engineer lower wages in areas of high unemployment so that these areas become more attractive to mobile capital.

Without entering into the problems of this crude neo-classical conception, this is the reasoning which led the Thatcher governments to claim that privatisation and trade union reform were part of its spatial policy repertoire to promote regional development, a precedent unique in the history of British regional policy. In other words, Thatcherism sought to emphasise the contribution which factors other than regional policy could play in securing inter-regional balance, factors like lower wages, decentralised bargaining and de-unionisation, etc.

This new emphasis on 'natural adjustment' owes much to the Conservative interpretation of US experience, where a combination of low wages and low union density has helped to attract mobile capital to the less prosperous 'sunbelt' states of the South (Morgan, 1985). The 'natural adjustment' argument surfaced again in 1994, as part of the Treasury's case for dismantling regional policy altogether, a move which seems to have been defeated by an alliance between the Department of Trade and Industry and the CBI (Wintour, 1994).

In stark contrast to the declining fortunes of regional policy, government expenditure on *urban* policy increased sharply in the 1980s. As we can see from Table 13.2, urban aid increased by 240 per cent in the decade to 1991/92, from £303 million to £1,031 million (local authorities would claim that these increases have been offset by cuts in main spending programmes). Within the urban policy package the most important expenditure item was spending on Urban Development Corporations (UDCs), which consumed well over 50 per cent of the total budget every year since 1989. Indeed, in 1990/91 the London Docklands Development Corporation alone accounted for 37 per cent of the entire budget, which suggests that the growth of *urban* policy at the expense of *regional* policy has been associated with a redistribution of expenditure from the peripheral regions in the north and the west to the south-east of England.

Table 13.2 Urban policy expenditure (1992/93 prices)

1981–82 (£m)	1991–92 (£m)	% change
303	1,031	(+240)

Source: Parliamentary answer (27 January 1993).

The flagship status of the UDC as a vehicle for urban renewal is not difficult to explain. The UDC had immense personal appeal for Mrs Thatcher because it was directly accountable to *central* government and a means of circumventing the influence of *local* government, an institution which was held in low esteem by the Thatcher governments. However, if UDCs dominated the urban policy agenda of the 1980s the post-Thatcherite era has spawned a number of new departures. A new phase was inaugurated at the end of 1990, when the government announced that it was extending its poll tax review to cover the structure and functions of local government. Among other things this review aimed 'to put the relationship between central and local government on a healthier footing, to replace conflict with partnerships, preferably within a widely accepted consensus about their proper roles' (Centre for Local Economic Strategies, 1991). It was in this context that the government decided to phase out the UDCs by 1995/96 and to create a new scheme in the shape of City Challenge.

Although City Challenge offered no new resources it embodied two important principles: first, that local authorities had to compete with each other to win resources and, second, that they were expected to forge partnerships within their localities, especially with the business sector and community organisations. Whatever the criticisms of the City Challenge philosophy – and there are several, not least what is to be done about the unsuccessful bidders – it nevertheless signalled a partial rehabilitation of the local authority role in urban renewal. Equally important, it created new incentives for collaborative action among public, private and voluntary sectors at the local level (Hambleton, 1991).

This emphasis on partnerships was reinforced when the government announced another new departure in the form of the Single Regeneration Budget (SRB), which brought some twenty separate programmes under a single budget as of April 1994, a move designed to create synergy out of a bewildering array of urban initiatives. To complement this new funding regime the government also introduced a potentially important institutional innovation in the shape of ten Integrated Regional Offices (IROs) in the English regions, the aim being to amalgamate the hitherto separate regional offices of four Whitehall departments, namely Trade and Industry, Employment, Environment and Transport.

The IROs are the government's concession to the argument that the lack of institutional coherence at the regional level in England has been a handicap in devising strategies for social and economic renewal. The IROs are also expected to act as facilitators for local partnerships, especially between local authorities and the local business sector, and to present SRB bids to ministers.

If partnership is to be the bedrock of urban renewal strategies in the 1990s it will have to be more than a rhetorical device, otherwise the UK will remain locked into the debilitating pattern of adversarial central-local government relations that did so much to vitiate urban initiatives in the 1980s. In 1993 the then Minister for Local Government set out a new vision for local authorities in the 1990s: among other things he said that local authorities should provide a leadership role in local economic development, and he urged them to 'promote, innovate and lead' (Redwood, 1993).

On the evidence to date, however, it is difficult to see how central government can expect local authorities to play the role of animateurs when local government as an institution has been so emasculated. The combination of draconian capping controls and compulsory competitive tendering is transforming local government into an 'adjunct' of the central state, so much so that local democracy itself is being compromised. The servile status of local government is most pronounced in the financial sphere, where over 80 per cent of its revenue is derived from central government (Audit Commission, 1993). The question arises as to whether real partnership between central and local government is possible in the face of such asymmetrical power.

Despite this inhospitable climate the local authority remains the key political institution at the local level. Even though its influence has been qualified by central controls on the one hand and by the burgeoning of local quangos on the other, the local authority can draw on resources – like democratic legitimacy and public accountability – which are denied to all other bodies at the local level. The new local governance system, in which the local authority is just one among many players, makes it that much more important for authorities to forge partnerships with cognate bodies in the local economy (e.g. TECs, Business Links, chambers of commerce, further education colleges, housing associations, etc.). However, the capacity to form partnerships is not uniform across the UK because so much depends on the political and institutional milieu at the sub-central level. In terms of its popular image Wales is thought to have a political and institutional culture which renders it well-placed to develop the kinds of partnership that are now deemed to be necessary for social and economic renewal. But what is the reality behind this image?

The Plight of the Valleys Today

With a population of 695,000, which amounts to 24 per cent of the population of Wales, the valleys of South Wales constitute a distinctive sub-region which has many of the social and economic problems associated with the inner city. Unlike cities, however, this sub-region is institutionally fragmented in that it embraces 16 district councils and 5 county councils, a situation which will change radically when the new unitary authorities are in place in 1996. This traditional pattern of local government is actually part of the problem in that it has allowed each valley to pursue its own parochial ends, to the detriment of the sub-region as a whole.

Whatever the nuances within the valleys they all share the same social and economic problems, many of which are associated with the decline of the coal industry. In employment terms the coal industry peaked in 1920, when 271,000 miners were employed in the South Wales coalfield. Ever since then the valleys have struggled to make the transition to a more diversified economy based on manufacturing and services. The economic structure of the valleys is biased towards manufacturing, which accounted for 33 per cent of the workforce in 1991, compared to 22 per cent in Wales as a whole.

In terms of socio-economic groups, manual workers accounted for 24.8 per cent of the workforce in 1991, employers and managers accounted for 9.2 per cent, while professionals accounted for just 2.2 per cent, whereas in Wales the proportions were 21.8 per cent, 11.8 per cent and 3.5 per cent respectively (Welsh Office, 1993b). In other words the valleys sub-region is more skewed towards low-skill manual occupations than is Wales as a whole.

All the evidence suggests that low skills attract low quality investment and this in turn begets low-skilled employment, the vicious circle in which the valleys sub-region finds itself today. If a low-skills economy is one of the underlying problems in the valleys, there are a number of other problems which are more palpable, namely:

- *unemployment*: Although the Welsh unemployment rate in September 1994 was not much higher than the UK rate (9.3 per cent against 9.1 per cent) the valley districts have unemployment rates way above this figure, reaching a high of nearly 25 per cent in the Cynon Valley.

- *wages and income*: In common with Wales as a whole the level of wages in the valleys has been on a downward trend throughout the 1980s, partly as a result of lower paid manufacturing and service jobs substituting for higher paid mining jobs. This has exacerbated the wider problem of low income in the valleys, where a high proportion of the population is on state benefits. This problem is most acute in Mid-Glamorgan, which has the lowest household income per head of any county in the UK.

- *population*: The valley communities have been losing population for many years: in the decade the Mid-Glamorgan valleys alone lost around 7,000 people, leaving the area with an ageing population profile.

- *environment*: The physical environment of the valleys still carries the scars of environmental degradation, a legacy bequeathed by the coal industry. The costs of this environmental legacy include the Aberfan disaster in 1966, when more than 100 children were killed when a loose coal tip engulfed their school.

- *housing*: The existing housing stock in the valleys, a large proportion of which is owner-occupied, is in a dire condition: once again some of the worst problems are to be found in Mid-Glamorgan, where one in ten houses is deemed unfit for human habitation, a figure which rises to 16.4 per cent in Cynon Valley and 15.3 per cent in the Rhondda Valley.

- *crime*: The growing level of crime is perhaps the most unnerving problem for communities which once prided themselves on a high degree of self-regulation. While reported crime increased by 20 per cent in England and Wales between 1985 and 1990, the rate of increase was 31 per cent in the central valleys over the same period (Morgan and Price, 1992).

What we have here is a noxious cocktail of social and economic problems, each of which feeds on the others, making it that much more difficult to tackle any one problem in isolation. Not surprisingly the valley communities dominate the official league table of urban deprivation in Wales, as measured by the Welsh Office Index of Deprivation. As we can see from Figure 13.1, of the fifty most deprived wards in Wales the overwhelming majority are to be found in the valleys sub-region. If we scale the ward data up to district level (as in Table 13.3, which shows the deprivation score), we get a league table of the ten most deprived districts in Wales.

Deprivation on this scale is not easily eradicated, and to suggest otherwise is utopian in the extreme. To be adequately addressed these problems require bold and imaginative initiatives at national, regional and local levels. Although there is only so much that can be achieved at the local level, localised initiatives were not particularly thick on the ground in the valleys until the launch of

Source: Welsh Office (1991 Census data) 1993.

Figure 13.1   Location of the fifty most deprived wards in Wales, 1993.

the Programme for the Valleys, and this is partly attributable to the political monopoly exercised by the Labour Party.

In terms of *parliamentary* politics the valleys sub-region has been an enduring bastion of British Labourism since the 1920s: the huge parliamentary majorities that we see in the valley seats today – 30,067 in Blaenau Gwent, 28,816 in the Rhondda, 26,713 in Merthyr, for example – have had the effect of stifling political debate. In other words the Labour Party's political hegemony has been so complete at the parliamentary level that local initiatives were deemed to be either unnecessary or at least subordinate to the goal of getting Labour into office at Westminster.

At the *local* political level, however, Labour's hegemony has begun to be challenged by Plaid Cymru, the Welsh national party. Starting from a low base Plaid Cymru appears to be winning over some of the younger elements of the electorate in the valleys. Although Plaid Cymru's support is in part

Table 13.3   The ten most deprived districts in Wales

Rank	District	Score
1	Rhondda	111.85
2	Rhymney Valley	92.55
3	Ogwr	81.47
4	Cardiff	78.39
5	Swansea	69.75
6	Cynon Valley	68.20
7	Taff Ely	66.48
8	Blaenau Gwent	62.24
9	Merthyr Tydfil	59.48
10	Neath	52.50

Source: Welsh Office.

a protest vote against the traditional ruling élite in local government, some commentators believe that a more fundamental process of political realignment is underway. Reflecting on recent local election results one expert argued that 'Plaid has won almost every valleys seat it has contested since the general election, fuelling beliefs expressed by academics that the "new-style" working class in the valleys is about to ditch the "old-style" Labour Party' (Betts, 1992).

Modest as it is, this local electoral shift has been enough to persuade some sections of the Welsh Labour Party that it needs to re-establish its credentials at the community level to enable it to design regeneration schemes with, rather than for, the valley communities (Morgan and Price, 1992).

Programme for the Valleys: The Partnership Approach

Because of its deeply entrenched Labourist traditions South Wales never embraced the free-market credo of Thatcherism. Nor, for that matter, did Peter Walker, whose political philosophy was much closer to the pre-Thatcher Tory tradition of Disraeli, Macmillan and Heath, a tradition which espoused an active role for government in social and economic renewal. When he became Secretary of State for Wales in June 1987, Walker brought to the Welsh Office a stature which had been missing since the Jim Griffiths era in the 1960s. Always eager to make his own mark, Walker launched the Programme for the Valleys in June 1988, exactly one year after taking office.

To some extent this initiative was designed to mollify the political lobby in the valleys, which was deeply concerned about the priority being accorded to the Cardiff Bay area. These fears were fuelled in March 1987, when the Cardiff Bay Development Corporation was formed to regenerate a 2,700 acre site in the south of the city. Since the 1930s the political lobby in the valleys has been acutely conscious of the fact that the centre of economic gravity within South Wales was shifting from the valleys to the coastal belt, where the physical geography is less of a constraint on economic development. Despite the natural advantages of the coastal belt the valley communities had always felt threatened by new projects to the south, so much so that they helped to kill the proposed

New Town at Llantrisant in the early 1970s, an episode which marked the high point in the political influence of the valleys. In the light of this intra-regional rivalry, the Valleys Programme was partly a sweetener, an attempt to pre-empt the valleys' opposition to the prestigious Cardiff Bay project.

Initially designed as a three-year initiative, the Programme for the Valleys was later extended into a five-year programme. At the launch of the programme Peter Walker said that the key aim was:

> to improve significantly the prosperity of the valleys of South Wales and the well-being of the people who live in them; to give people a new confidence in the future of their valleys as places in which to live and work and to instill in people elsewhere a new perception of the area as a place worth visiting and investing in.
>
> (Welsh Office, 1988)

More concretely the programme aimed to reduce unemployment by up to 30,000, improve education and training provision, remove environmental degradation, enhance the housing stock and raise the quality of social and community life. Another aim which deserves to be mentioned – an aim which was given added prominence by David Hunt, Walker's successor – was the promotion of partnerships between central and local government, public and private sectors, employers and unions, in short between all the parties engaged in social and economic renewal.

In total a sum of £800 million was originally committed to the Valleys Programme, and the breakdown of expenditure is shown in Table 13.4. Two points need to be made about the budget. First, the actual expenditure for the five-year period was £751 million rather than £800 million, a shortfall which can be attributed to the fact that some of the schemes (like the regional

Table 13.4   Five year expenditure profile of the Valleys Programme 1988–1993

	£ million
Urban Programme	92
Urban Development Programme	25
Roads Expenditure – Transport Grant allocation to LAs	81
Roads Expenditure – Trunk Roads	12
Welsh Office Support for Industry – RDG	35
Welsh Office Support for Industry – RSA	55
Welsh Office Support for Industry – REG	1
WDA factory building	95
WDA Business Development	4
WDA Investment	7
WDA Urban Development	19
WDA Derelict Land Clearance	77
WDA Environmental Improvements	5
Employment Department programmes	243
Total for Programme elements	751

Source : Welsh Office.

industrial grant for example) were demand-led and demand tapered off during the recession. Second, it is also clear that the programme involved little or no additional resources over and above what was already committed to pre-existing schemes (Rees, 1989). As we can see from Table 13.4, many of the schemes under the Valleys Programme are available in most of the UK's assisted areas.

In material terms then the main criticism of Walker's initiative was that it was essentially a marketing exercise: that is to say it brought together what had hitherto been a loose amalgam of schemes and presented the package as a wholly new programme. Even so, the packaging process recognised, in principle at least, the critical importance of having an *integrated* approach to urban renewal.

As the five-year Valleys Programme ended in March 1993, what has been achieved during this period? According to the Welsh Office the key achievements include:

- nearly £700 million of additional private sector investment, involving 24,000 new and safeguarded jobs
- 2.6 million square feet of new industrial floorspace
- over 2,000 acres of derelict land cleared by the WDA in what is said to be the largest clearance programme in Europe
- more than 7,000 homes improved and refurbished
- the most successful of all the National Garden Festivals, at Ebbw Vale
- new arts facilities, including 'Valleys Live 92', a major festival to celebrate the arts (Welsh Office, 1993)

Evaluating the impact of the Valleys Programme is beset by all the standard methodological problems, like the counterfactual problem of what might have happened in the absence of the programme and what causal significance we can attribute to the programme when it was just one among many factors influencing social and economic trends in the sub-region (DoE, 1994c). Evaluation is further compounded by the fact that the programme period coincided with one of the deepest recessions of the post-war era.

Having made these qualifications it is still necessary to make a judgement about the probable impact of the programme. As regards the key aim of the programme – which was to reduce the *unemployment rate* in the valleys – the best that can be said is that there was some marginal improvement. As we can see from Table 13.5, the unemployment rate fell slightly in the valleys sub-region over the programme period, so that while the area fared better than the UK as a whole, it lagged behind the improvement in Wales, where there was no programme!

Table 13.5 Comparative unemployment rates: 1988 and 1993

	Valleys	Wales	UK
June 1988	14.0	11.2	9.2
June 1993	13.8	9.9	10.2

Source: Employment Department.

If we take *wage levels*, another key to improving prosperity, then it is clear that the programme failed to make any impression on the growing gap between local and national wage levels. For example, average weekly male earnings in Mid-Glamorgan, which accounts for some 60 per cent of the population of the valleys, continued to fall throughout the 1980s, from over 99 per cent of the national average at the start of the decade to 88.8 per cent in 1992. On the economic front in general the official evaluation found that 92 per cent of programme managers felt that the range of economic development initiatives was insufficient to address the economic problems of the valleys (Victor Hausner and Associates, 1993).

Aside from its economic development objectives the programme sought to enhance education and training facilities, improve the environment, boost the arts and raise the quality of health, housing and transport provision. While some progress was reported in each of these spheres, the overall conclusion was that 'the Programme has helped to arrest the deterioration and decline of economic and social conditions in the South Wales Valleys' (Victor Hausner and Associates, 1993).

Aside from resource constraints the main weakness of the Valleys Programme was on the management side. For example:

- the programme was designed as a series of separate initiatives which lacked clearly defined strategic and operational objectives
- the lack of central co-ordination within the Welsh Office and the poor communication links between the latter, the WDA and the local authorities inhibited the effectiveness of the programme
- the local authorities, who were responsible for delivering parts of the programme, were generally critical of the lack of Welsh Office guidance
- inadequate strategic and operational objectives meant that the quality of monitoring information left much to be desired
- there was little evidence to suggest that the information that was collected was used to improve the performance of the programme, rather this information was used simply to adjust target spending
- there was no systematic dissemination of good practice and supportive technical assistance and training for the parties involved in the programme, and this is essential to building managerial capacity at all levels (Victor Hausner and Associates, 1993).

On the positive side the programme has certainly raised the profile of the area, signalling to public and private sectors the new political priority accorded to regenerating the valley communities. The evaluation exercise found a good deal of support from all bodies in the sub-region for the partnership approach to socio-economic renewal, where the key innovation was said to be the urban joint-ventures between local authorities and the WDA. Indeed, the WDA was perceived to be a major asset, not least because it gave the Welsh Office an implementation capability not matched by central government outside Wales. Even so, this favourable institutional capacity was not enough to offset weaknesses in programme design.

The weaknesses in programme design were further compounded by the fact that at least three central government policies ran counter to the aims and aspirations of the Valleys Programme. First, regional policy expenditure in the programme area was reduced by some 50 per cent in real terms during the first three years of the programme. Second, the programme was unable to prevent the attrition in training places in the valleys, a consequence of cuts in TEC budgets. And third, the financial position of local authorities in the valleys is as weak as ever, again because of central government policies (Morgan, 1992). What this means is that central government policies must be better synchronised with, and more supportive of, local strategies for urban renewal.

Notwithstanding all these problems it is worth saying that one of the most important − but least tangible − achievements of the Valleys Programme is that it firmly established the principle of partnership, or collaborative action, as a key mechanism for the regeneration of urban areas. Even the beleaguered local authorities in the valleys have gone out of their way to welcome 'the new spirit of partnership' which was induced by the programme (Roberts, 1991). However, if this spirit is to be sustained, and if the partnership approach is to deliver tangible benefits, then new governance structures will have to be created to manage the process of social and economic renewal, an issue which I address in the following section.

Conscious of the fact that a five-year programme cannot reverse sixty years of decline, and anxious as it is to maintain the momentum, the Welsh Office launched another five-year Valleys Programme in April 1993 at a total cost of well over £1 billion in public sector investment (Welsh Office, 1993a). Furthermore, John Redwood, the new Secretary of State for Wales, expects that this will lever in an additional £1 billion in private sector investment over the next five years. While the details of the second Valleys Programme are still being elaborated, the principles on which it will operate were outlined by David Hunt before he left the Welsh Office (Hunt, 1993a). The three most important principles were:

- Firstly, to deepen and extend the *partnership* approach to urban regeneration by harnessing the resources of the WDA. In recent years the Agency has assumed an increasingly important role in urban renewal, so much so that its urban budget has been increased from just £8 million in 1989/90 to £34 million in 1993/94.

- Secondly, to secure a more *co-ordinated* approach to urban renewal by integrating the full range of services, like road and rail infrastructure, land reclamation, property development, business services, training and community development schemes.

- Thirdly, to move towards a more *decentralised* model of urban regeneration by tapping local initiative to a much greater degree than in the past. Belatedly, the Welsh Office now realises the crucial significance of building *local capacity* by raising the professional, managerial and networking skills of each party in the renewal process, without which the decentralised model is doomed to failure.

These are the key principles of the Welsh Office's urban renewal strategy for the 1990s, each of which has been endorsed by the main political parties in Wales (Labour Party, 1993; Plaid Cymru, 1993). With this degree of consensus there is a good chance that the programme will at least achieve the necessary continuity, an ingredient that has been missing in the past. However, two other ingredients will also be necessary, namely, more effective and accountable governance structures on the one hand and more supportive central government policies on the other.

From Partnerships to Governance Structures

The notion of 'partnership' can so easily degenerate into a meaningless slogan, which was the fate of 'flexibility' in the 1980s. In the most general terms partnership might be defined as the disposition to collaborate to achieve mutually beneficial ends, a definition which is broad enough to embrace anything from the informal exchange of information to more formal joint-ventures, in which resources and expertise are pooled so as to achieve a clearly defined social or economic objective. Imprecise as it is, the notion of partnership is a key element in network theories, which eschew with equal vigour the dirigiste approach to development, which exaggerates the capacity of the state, and the neo-liberal approach, with its naïve faith in market-based solutions (Cooke and Morgan, 1993).

We are now witnessing a veritable explosion of public-private partnerships in the UK, in such fields as urban renewal, local economic development, training provision and environmental improvement. To a large extent this phenomenon is being driven by two pressures: first, funding regulations are such that a 'partnership approach' is required to access financial aid from both London and Brussels and, second, there is a growing recognition that no single organisation has the expertise or the resources to offer credible solutions on its own, hence the need to collaborate.

Whatever the activity the local partnership raises a number of questions, not least about its effectiveness as a vehicle for regeneration and its accountability to the community in which it operates. These issues lie at the heart of the debate about how local economic development should be managed at a time when many of the traditional functions of local government have been transferred to 'non-elected local states' like TECs and UDCs (Peck and Tickell, 1992). In other words the growth of local partnerships needs to be situated in the context of the debate on new governance structures. To examine these issues in more detail this section explores the question of governance structures at three different spatial scales: at the local level (Cynon Valley), the sub-regional level (the Valleys) and the Wales-wide level.

At the *local* level the Cynon Valley is as good a focus as any because of all the local authority districts in the Valleys none brings more energy and commitment to the task of urban renewal than Cynon Valley Borough Council (CVBC). Indeed, the Welsh Office perceives CVBC as something of an exemplar, one which other local authorities ought to emulate. One measure of its success is that it has won a larger share of urban aid than most of its neighbouring districts, even though the latter have a more deprived status (Morgan, 1992). What also distinguishes CVBC is that it has been in the forefront in forming partnerships in skills provision, housing, business development and urban renewal. Of all these partnerships the urban joint-venture with the WDA is probably the most innovative and the most important.

The urban joint-venture is part of the WDA's Urban Development Wales programme, in which some twenty-eight towns have been targeted for varying levels of investment. Within this programme the urban joint-venture enjoys the highest status, signifying the areas where the WDA intends to make its biggest commitments. In targeting its joint-venture partners the WDA uses two criteria: first, the areas where the prospects for levering in private investment are best and, second, districts where the managerial calibre of the local authority is deemed to be highest. To date the WDA has forged a number of urban joint-ventures throughout Wales, but just two of these are in the valleys sub-region, in Merthyr and Cynon Valley (Welsh Development Agency, 1992).

Launched in 1991 the urban joint-venture in the Cynon Valley is a tripartite partnership between CVBC, Mid-Glamorgan County Council and the WDA. Among other things this partnership aims to create a number of high grade business sites, to redevelop the two main towns (Aberdare and Mountain Ash), to enhance the infrastructure of the valley, to promote a 'green corridor' through environmental improvement projects and to support local training schemes, like the new Technology Centre, which caters for electronics-related skills. With a budget of nearly £20 million, most of which has been borne by the WDA, the joint-venture has levered in some £35 million in private sector investment.

In governance terms the key mechanism is the joint-venture board, which consists of representatives from each of the three parties. Decisions taken by the board are referred back to individual Council committees and the minutes of board meetings are circulated at full Council meetings for comment and approval. Apart from a controversial land deal, in which the board was accused of favouring one private company over another, the main teething troubles have stemmed from a clash of cultures: the WDA sets a high premium on working to strict commercial deadlines, while the local authorities tend to work at a slower pace, not least because, being publicly accountable, they have to win approval for their actions. Nevertheless, the general consensus seems to be that this is a creative tension in the sense that the WDA encourages the local authorities to be more innovative, while the latter seek to impress upon the WDA the virtues of community support. In the words of the local Labour Group:

> The task of regeneration could not possibly be completed by underfunded and disempowered local Councils. The need, therefore, for the involvement of a national development agency is unarguable. The WDA has brought specialisms and expertise to the Joint Venture that could not have been provided by local authorities . . . . The Cynon Valley Urban Joint Venture has shown that central government need not sacrifice local democracy in order to carry out the task of regenerating communities.
>
> (Cynon Valley Borough Council Labour Group, 1994)

Whatever the problems with this approach they are as nothing compared with the defects of the UDC model, in which local authorities are often sidelined and where there is little or no local public accountability. The fact that the urban joint-venture model can be shown to be both effective *and* accountable proved to be a telling argument against the creation of a UDC for the valleys, an idea which was at one time under consideration at the Welsh Office.

At the *sub-regional* level we have a patchwork quilt of bodies, none of which has sole responsibility for the regeneration of the valleys as whole. It was this lack of institutional focus which led the Welsh Office to toy with the idea of a UDC with a brief to focus on the valleys as a sub-region. Although the idea was rejected, the problem remains. Indeed, one of the weaknesses of the first Valleys Programme was the absence of a sub-regional forum in which valleys-based organisations could compare their experiences, co-ordinate their activities and, most important of all, offer an alternative perspective to the Welsh Office as regards the effects of the programme.

To overcome these problems the local authorities took the initiative to create the Valleys Forum, which consists of representatives of all the relevant organisations in the sub-region (e.g. local authorities, TECs, FE colleges, WDA and community development groups). At one level the Forum provides a mechanism for these bodies to share their experiences and disseminate best practice, at another level it will monitor the new Valleys Programme and act as an interlocutor with the Welsh Office on the design and delivery of regeneration projects. When the new programme was launched the then Secretary of State said:

> I particularly welcome the creation of a Valleys Forum, driven by local people, to give a unified voice to all the organisations which are active in the valleys.
>
> (Hunt, 1993b)

In governance terms, however, the Valleys Forum is just a loose organisation with a voluntary membership base. Most important of all is the fact that the Forum lacks any formal authority, hence there is nothing to prevent the Welsh Office from simply ignoring its recommendations. While the Forum commands a good deal of moral authority, on account of its civic standing, it has no power to shape the activities of the key institutions in the sub-region, namely, the Welsh Office, the WDA and the TECs.

To compensate for the lack of governance capacity at the local and sub-regional levels the focus has now moved to the *Wales-wide* level. Two factors have propelled the governance question to the top of the political agenda in Wales: first, the widespread concern about the 'democratic deficit' associated with the proliferation of quangos and, second, the growing demand for a more coherent regeneration strategy.

The burgeoning of quangos (quasi-autonomous non-governmental organisations) in the UK has been little short of remarkable, and nowhere more so than in Wales. Although the number of quangos in the UK has actually fallen, the amount of public expenditure at their disposal has increased from £13.9 billion in 1979 to over £40 billion today. In Wales, the number of quangos has actually increased, from around 40 in 1979 to well over 80 in the early 1990s. The influence of quangos in Wales is evident from the fact that, in total, they account for over £2 billion, which is more than a third of the total Welsh Office budget, and they employ over 57,000 people (Morgan and Roberts, 1993; Morgan, 1994a). To counter the quango-state, which is unelected by and unaccountable to the Welsh electorate, all the political parties in Wales, with the exception of the Conservative Party, are now in favour of a directly elected Parliament for Wales. Hence the main rationale for a Parliament is that it would subject the Welsh Office and its network of quangos to democratic scrutiny (Osmond, 1994).

Aside from the issue of democratic governance there is also the question of how a more devolved governance system could help to stimulate a more robust strategy for social and economic renewal. The Welsh Office is now responsible for a wide array of policies, in such fields as education, training, enterprise support, urban and regional development and transport, etc. In other words it is ideally placed to stimulate local partnerships and orchestrate the disparate sources of expertise in Wales. Being an outpost of central government, however, the Welsh Office has been criticised for its narrow conception of partnerships. In a review of the partnership framework in South Wales, for example, the European Commission concluded by saying:

> Many of the partners see partnership in a *horizontal* sense in which equal partners work towards common ends. Central government has tended to see partnership in *vertical* terms in which the Welsh Office plays the decisive role with any alternative model viewed as unacceptable . . . . Working relationships in industrial South Wales, although reasonably good, are limited in scope.
>
> (European Commission, 1991)

More recently the new political masters in the Welsh Office have begun to question whether there is a need for a WDA role in urban renewal and business services (Morgan, 1994). Under the direction of John Redwood, a prominent member of the new right, the Welsh Office is perceived to be less than enthusiastic about having a robust development agency in Wales. What this means is that the political consensus which has developed in Wales, a consensus which included John Redwood's Conservative predecessors at the Welsh Office, is beginning to unravel because of the neo-liberal priorities of a governing party which, in Welsh terms, is a *minority* party, with just 6 of the 38 seats in Wales.

A more democratic governance system, in the form of an elected Welsh Parliament, would be far more sympathetic to the horizontal model of partnership, a model which is better able to harness the full potential of the economic development community in Wales, a model which is also more democratic in character. The current Welsh Office strategy, geared as it is towards the more restricted vertical model of partnership, is a symptom of the over-centralised and undemocratic nature of government in Britain today (Lewis, 1992; Morgan, 1994).

However, the trend of political opinion in Wales is now running in favour of a directly elected Parliament. What informs the Parliament for Wales campaign is the belief that decentralised governance structures, which are locally attuned and publicly accountable, are more effective, more democratic and therefore more sustainable than the centralised structures we have at present. Creating stronger and more inclusive governance structures at the sub-central level is no easy task in a unitary-minded country like Britain. On the other hand the status quo no longer seems to be a viable option for the future in Wales.

Conclusions

The main aim of this chapter has been to examine the key urban regeneration initiative in Wales against the background of UK territorial policy on the one hand and sub-central governance structures on the other. Among other things it was argued that social and economic renewal depends in part on institutional capacity at the sub-central level, and that such capacity tends to be weak in the UK compared to the more dynamic regions of the European Union. Within the UK, however, Wales enjoys a measure of institutional autonomy which is denied to the English regions. Modest as it is, such devolved power creates the potential for a more coherent and integrated approach to urban regeneration than is possible in England. Indeed, it seems highly unlikely that an initiative on the scale of the Valleys Programme would have emerged in Wales had the latter been a mere replica of the English regions.

While the Welsh Office makes great claims on behalf of the Valleys Programme – the key urban regeneration initiative in Wales – the record suggests that these claims are exaggerated. The most charitable thing we can say is that the programme is necessary, but in no way sufficient, to regenerate the valleys, a task that is clearly beyond the scope of a five-year programme. Aside from the charge that the programme involved no additional resources, the most important criticism to be made concerns the management of the programme, a point underlined by the official evaluation study. In *design* terms the programme was little more than a series of separate initiatives, hence there was little or no synergy between the projects. The *implementation* phase was marred by the lack of central co-ordination from the Welsh Office and by inadequate communications among the partners. Finally, *monitoring* information was poor and little effort was made to disseminate best practice through the programme area.

In other words, even though Wales is better endowed with institutional capacity than the English regions, this potential was not exploited. Indeed, the Valleys Programme suffered from many of the design and delivery weaknesses which were identified in the management of urban policy initiatives in England (DoE, 1994). Furthermore, localities in both Wales and England suffered from another key problem: that the grain of central government policies ran counter to the aims of locally designed initiatives, a sobering reminder of the fact that while little victories can be achieved at the local level, central policies are too intrusive, too influential and ultimately too powerful a force to be overcome at the local level. Without a supportive macro-environment, little victories are fragile achievements.

On the positive side, however, it must be said that the Valleys Programme has induced a new spirit of partnership in the sub-region, and this will need to be sustained under the second phase of the programme. One of the most innovative partnerships within the programme were the urban joint-ventures between the WDA and the local authorities. This partnership model, which is designed to marry the technical expertise of the WDA with the local knowledge of elected councils, is not possible in the English regions because the latter have no equivalent of the WDA.

Although the urban joint-venture model can rightly claim to be more accountable, in governance terms, than the UDC model, it is nevertheless too limited in scale and scope to have a wider impact; it is certainly unable to influence the decisions of the unelected quangos which dominate the economic development scene in Wales. Herein lies the significance of the Parliament for Wales campaign because what animates this cross-party movement is the conviction that new governance structures are necessary not just to redress the democratic deficit, but to stimulate more innovative partnerships for social and economic renewal. With Wales and Scotland making ever more vigorous demands for *democratic* devolution, it may be that this pressure will trigger similar demands from the English regions. Whatever the future holds we have clearly passed the point where the current Whitehall-dominated system could credibly claim to be the most innovative, or the most accountable, form of governance in this country.

14

URBAN POLICY IN ENGLAND AND FRANCE

Howard Green and Philip Booth

Introduction

For the past ten years the problems associated with urban living and their unequal incidence on city dwellers have been a preoccupation for governments throughout Europe. In Britain there was already by the late 1960s a recognition of an 'inner city problem'. In France the same concern for an apparently similar set of interconnected problems has focused on the city fringes. This chapter represents a preliminary stage in a comparative study of two policy programmes: City Challenge in Britain and Contrat de Ville in France which appear to propose strategic action to combat the problems in much the same way. The chapter concentrates on the French approach to urban policy represented in Contrat de Ville, and presents in some detail the contract for Lille. The intention is not only to describe but to raise questions about the policy process that might equally be applied to Britain.

The Problems of Comparing Urban Policy

Making a comparison between urban policy initiatives in the two countries is on the surface appealing. On the one hand, both countries appear to be facing a similar set of urban problems which are social and economic in origin and have a direct impact on the well-being of certain sectors of their urban populations. On the other, both countries have embarked at more or less the same time on a policy programme designed to make a fresh start, which is both strategic and inter-sectoral. Contrat de Ville in France may be seen in the same context as City Challenge in England and Wales as an innovative approach to urban regeneration. But to propose a comparative study is to face immediately a series of intractable conceptual and methodological problems. The apparently beguiling similarities dissolve on closer inspection into a host of differences, as other research has demonstrated (Davies, 1980; Jacquier, 1991).

We would nevertheless argue that there is a role for comparative study, not in terms of the transfer of policy, but in terms of the process whereby policy is articulated and then applied. This argument rests in the belief that the way in which problems are dealt with is determined in large measure by the political framework and the pattern of decision-making within the country under study. The value of looking at this process in a different culture is to highlight what is all too readily taken for granted in one's own (Wolman, 1993).

In considering the policy process surrounding Contrat de Ville, there are essentially questions at two levels. At one level we need to ask if the initiatives are in fact the radical departure that they were presented as being, and if so, in what way, how the initiatives were conceived and how they were then applied in particular places. At another level, we need to look below the surface of the programmes themselves. Here we are interested in three elements which can be conceptualised as mechanisms, relationships and agencies.

The Contrat de Ville may be seen as a mechanism by which particular policy objectives can be articulated and achieved. The analysis will assess the way this particular instrument provides the mechanism to bring together the various interest groups and individual sectoral policies into a comprehensive policy for urban areas. It will reflect on the language that is used to express the policy content and the values that the language implies.

Relationships are increasingly emphasised in urban policy development and are frequently discussed in terms of partnership development (Whitney and Haughton, 1990). Partnerships are not new to French policy-making and implementation. The Sociétés d'Economie Mixte, for example, date from the 1950s (Ministère de l'Intérieur et de l'Aménagement du Territoire, 1993). Partnership has come to include less formal structures which bring together other agencies, public, private and community based, and in particular the various tiers of local administration. The extent to which the Contrat de Ville allows or encourages the development of relationships between the various partners is of particular interest.

Finally in our framework is the consideration of agencies. We have noted elsewhere that the proliferation of agencies has been a characteristic of policy process since the 1980s (Green and Paris, 1992). An understanding of the role and approach of the many agencies involved in the urban policy process will greatly aid effective policy development and implementation. Such understanding will be concerned with issues of leadership and co-ordination in the policy process. In the context of Contrat de Ville, the analysis must consider the agencies involved in the development of the different aspects of policy and the way in which agencies are controlled and co-ordinated.

The three concepts of mechanisms, relationships and agencies will form the focus of our analysis. But before we can proceed to this kind of analysis, some description of the Contrat de Ville is necessary.

French Urban Policy and Contrat de Ville

France's concern for what might loosely be called 'inner city problems' is rather more recent than Britain's. The 1960s and 1970s were dominated by programmes for building housing that was mainly in the form of tower or slab blocks and to a large extent on peripheral sites. Nevertheless by the late 1970s problems of unemployment, social exclusion and restlessness were beginning to emerge in precisely those major developments that had been the result of urban policy a decade earlier. These difficulties were given dramatic emphasis by the riots at les Minguettes in the Lyon suburbs in 1981. Though not as severe as the riots in Brixton and Toxteth, the impact on French public policy was at least as dramatic, and led to the programme known as Développement Social des Quartiers (DSQ), started in 1984. This aimed to tackle the social causes of difficulties not only in peripheral estates but also in older urban areas and by 1992, 271 quartiers had been subject to DSQ agreements under this programme (Commissariat Général du Plan, 1992). Concurrently, the Banlieues 1989 programme aimed to find innovative solutions to the physical problems of high-rise estates.

The programmes did not, however, seem to address the whole problem. Those charged with implementing policy soon found that the problems of the quartiers en difficulté could not be resolved without reference to the urban area in which they were located. Problems of physical isolation, employment and social exclusion could only be tackled at the level of conurbation. Contrat de Ville (CDV) was born of this realisation. The intention was not to do away with DSQ but to integrate them within a global strategy which would commit central and local government to a programme of action. FF263 million was budgeted for the programme with the hope that up to 200 contracts might be signed in the first five years.

The first steps towards implementing the policy were slow. The Délégation Interministérielle à la Ville et à l'Action Sociale (DIV) set up to handle the new urban policy recognise now the difficulties that they faced in achieving the original targets. By late 1991 only thirteen contracts were signed or ready to be signed. These thirteen contracts represented an enormous diversity of types. At one end of the scale was St-Dié-des-Vosges consisting of a single unit of local government with 35,000 inhabitants. At the other was the Contrat for Seine-St-Denis covering a whole département with a population of 1.3 million (Booth and Green, 1993).

Events, however, have given the policy impetus. The first was further rioting, at Vaulx-en-Velin in Lyon and Mantes-la-Jolie in the Ile-de-France, in September 1990. This led to the creation of a Minister for Towns and an emphasis on CDV as the mechanism for ensuring that the quartiers were dealt with effectively. The second has been the vigorous underwriting of his predecessors' policy by Balladur after April 1993. Thus Simone Veil was appointed second-in-command and Minister for Towns, and Contrat de Ville is to become the sole contract for urban problems between state and urban areas.

In terms of the mechanism used, the role of the contract requires some exploration. As Méjean (1992) has shown, the contract as a means of articulating policy goes back some twenty years, and was initially seen as a way of ensuring the implementation of state strategy in the regions with decentralisation of power to local government from 1982 onwards.

Contracts came to have an increasingly important role in defining responsibilities of central government and local authorities alike. Most important of these contracts was the Contrat de Plan in which the intentions of the national five-year plans are translated into agreed action for each of the twenty-two regions.

CDV falls, therefore, into an administrative practice of some standing. We would argue that the significance of the contract is to make good the difficulties caused by the fragmentation of local authorities (France has more than 36,500 communes, the base units of local authority) and what the Commissariat Général du Plan (1992, p. 37) calls the 'entanglement of responsibilities'. Whether it achieves that end is open to question.

The second comment concerns the role of DIV and the insistence on inter-sectoral working. This must be understood not only in terms of the conservatism of, for example, the Ministry of Finance, but also of a conflict with the Délégation à l'Aménagement du Territoire et à l'Action Régionale (DATAR), itself set up in the 1960s with an integrative role in public policy-making. Because DATAR has been fiercely defensive of its role in economic development, economic policy and associated initiatives have so far occupied a relatively low place in CDV. Setting up DIV has possibly been self-defeating (Guichard, 1993).

In terms of the language two concepts are worth highlighting. The first is solidarité, with its implications of partnership and pulling together. This is not, however, to do with public/private partnership that preoccupies British policy-makers. The reference is rather to the centripetal forces that result from the fragmentation of local government. In particular solidarité refers to the difficulties faced by small, poorly resourced communes which may nevertheless have a disproportionate share of a conurbation's problems.

The second key concept is exclusion. Although the idea that parts of the population are being marginalised by their economic and social condition is not specifically French, in France the term derives particular force from its ethnic minority population. Exclusion is thus to be understood in the context of the debate about nationality, current at least for a decade.

We are now in a position to enquire in more detail into how the explicit objectives of the programme have been met and what effects the institutional and conceptual framework have had in achieving the objectives in practice.

The Contrat d'Agglomération de Lille

The use of case study in the exploration of a policy initiative such as the Contrat de Ville inevitably raises the usual methodological difficulties associated with the approach. It is particularly problematic in this case because of the heterogeneity of the towns and cities involved in the specific policy. Those already approved include areas of very diverse histories, economic structures, current economic problems, demographic characteristics and sizes and local administrative structures. Each of these characteristics has significance for the nature of the individual contract and the processes adopted in implementation. On the other hand, if we are interested in understanding process in particular, case study and the choice of the actual case can perhaps be approached in a more pragmatic way.

The choice of the metropole, Lille, Roubaix, Tourcoing, was made because the authors had long experience of working in the area and had built up a long-standing relationship with both academics and professionals working in the field of urban development. Access to documents and professional discussion was facilitated by this existing relationship.

The case study will allow many of the key aspects of the policy to be identified and assessed. The study of Lille will allow generalisations to be made in some areas of policy, in others it will not. For example, the history and economic structure of Lille sets it aside from other towns and cities in terms of specific problems and hence the associated policies and initiatives. Equally the administrative structure suggests that the processes associated with intercommunality will be different, although there is a long tradition of intercommunal working.

There will, on the other hand, be areas of similarity which will allow greater understanding of the French process. The relationship between the DIV and the locality and the transferability of nationally developed theoretical schemes to individual localities have been identified as key elements throughout the policy initiative. Similarly transferable is the relationship between DIV and DATAR in terms of local economic policy.

The nature of relationships within the Contrat for Lille will be of general concern, between for example the various technical agencies and departments, between the elected representatives and between the various administrative layers with their respective and closely guarded responsibilities, leading as in many other areas of French development to the emphasis placed on 'contractualisation'.

Lille is the fourth largest conurbation in France with a population in excess of 1,000,000. The current administrative authority, the Communauté Urbaine de Lille (CUDL), was created in 1966 as part of the policy of metropoles d'équilibre (Moreau, 1989). The authority is made up of 87 individual communes, each with its own mayor and responsibilities. In terms of urban structure, it is made up of three towns, Lille, Roubaix and Tourcoing, each with its own tradition, political imperatives and diverse problems.

The area which the contrat covers is included in that of the Schéma Directeur (SD), the strategic land-use planning document, which is currently being revised, and the POS (Plan d'Occupation des Sols). It is a complementary document to both the POS and SD, dealing with social and housing policy matters. The three are seen as part of a policy process which add up to a Projet Metropolitain. In this context the Lille contract has avoided some of the problems identified by the Commissariat Général du Plan (1992).

The present economic structure of the communauté urbaine is the result of the structural changes which have taken place in the 1980s, in a region associated with traditional industries, coal mining, steel, chemicals and textiles. Although growing, the service sector is under-represented. Having suffered significant decline, manufacturing still plays an important role in the local economy. Unemployment is above national average.

Because of its position within northern Europe, significant developments are currently in progress, not least those associated with the TGV and channel tunnel and north European links. The economy is a key issue in the development of the metropole.

In part because of its economic structure and associated problems, the region has a well-developed tradition of intercommunal working and experience of innovation and experimentation in matters of planning and development. It was, for example, at the forefront of the contractualisation Contrat de Plan, État/Région, was involved in the pilot stage of the DSQ and now has a well-developed DSQ programme. The notion of CDV is therefore not new to the CUDL. It is however questionable whether the CUDL is too large geographically and too complex administratively for successful implementation of the contract.

The DIV had initially wanted to agree three individual contracts with Lille, Roubaix and Tourcoing because of the complexity of policy implementation across 87 communes. This proposal was resisted because of continuing political commitment to policy development across the conurbation. The Contrat covers the whole conurbation and offers the opportunity to relate areas of neglect and degradation with those of development and affluence for the first time strategically.

The Contrat

The development of the Contrat d'Agglomération, the title of the Lille initiative, began soon after the announcement of the programme by Michel Rocard in 1989. The final Contrat between the three key partners, the state, the region and the Communauté Urbaine, was signed in January 1992. The development of the Contrat was a slow process even in an area with experience and commitment to joint working.

The Contrat is spelt out in a 33-page document (Communauté Urbaine de Lille, 1992a). The document details the eight objectives (see Table 14.1), the associated themes and the articles which define the precise contractual relationship. A separate document (Communauté Urbaine de Lille, 1992b), presents the details of the proposals, costs and funding arrangements.

Table 14.1 highlights the importance placed on social issues and the quality of urban life; combined, they represent over 75 per cent of the proposed funding, and illustrate the limited attention given to economic development. The initiatives within the Contrat are remarkably similar to those of the DSQ programme (Ville de Lille, 1989). The key differentiating feature is the scale, both geographically and financially, of the proposals.

Partnerships are an important element of the Contrat. The partners vary with the individual projects as one would expect. However, analysis of projects confirms that the partners are generally from state or state-funded organisations, rather than the private and voluntary sectors as is increasingly observed in the UK. For the majority of actions the partnership is made up of the state, the communauté urbaine, the communes, the region, the department and organisations such as Fonds d'Action Sociale (FAS), Caisse d'Allocations Familiales (CAF) and SNCF. The task is one of co-ordinating the actions of the existing state agencies, each with different responsibilities, within a strategic and agreed framework.

The contributions of the partners to funding provides an insight into the structure of the partnerships and the relative importance of each (see Table 14.2).

Each of the partners has its own responsibilities which generally are undertaken independently. The CDV brings these partners into an agreed strategic framework. Here we see a significant development to the DSQ in which the quartier policy was developed in isolation.

In terms of overall financial commitment, the communes contribute 53 per cent of the total amount, the state over 38 per cent. This balance reflects the pattern of main programme funding in France with its emphasis on the state.

Table 14.1   Contrat d'Agglomération: objectives and topics (in thousand francs)

Objective 1: marginalised groups and people at risk	
(a) Housing	
(b) Young children	
(c) Promotion and support for educational success	
(d) The handicapped	
(e) Gypsies	
(f) Prevention of delinquency	
(g) Developing access to cultural activities	
(h) Health	
Total cost:	605,742
Objective 2: getting people back into the mainstream economy or into training	
Total cost:	34,163
Objective 3: the environment and quality of urban areas	
(a) Urban restructuring	
(b) Urban upgrading	
(c) Environment and pollution	
Total cost:	519,700
Objective 4: the town and the university	
(a) University developments	
(b) Student accommodation	
Total cost:	15,150
Objective 5: public transport	
Total cost:	8,000
Objective 6: international aspects	
Total cost:	200
Objective 7: new communication technologies	
Total cost:	840
Objective 8: the management of the Contrat d'Agglomération	
Total cost:	8,150
Grand total	1,490,695

Source: Communauté Urbaine de Lille (1992a).

It also provides us with an insight into the importance of the communes, particularly when working together.

The process of development and implementation is the responsibility of a group or committee each with specific responsibilities. The structure is complex and is one which places a heavy burden on the Agence de Développement et d'Urbanisme in maintaining strategic direction. This organisation operates in a quasi-consultancy role outside the strict administrative structure, and is funded jointly by the state, the département and CUDL. It is charged with the overall initiation of projects with the various partners, and the structuring of funding packages including grants.

Table 14.2   Partners and funding contributions

Funding source	Contribution (thousand francs)
State	572,277
Contrat de Plan and European Credits	30,905
Region	13,990
Département	22,248
Communes and CUDL	798,183
Other European	3,200
Other partners	49,892

Source: Communauté Urbaine de Lille (1992b).

A more detailed analysis of the housing topic within Objective 1 allows us to identify at a finer grain some of the key elements of the Contrat. Housing policy is expressed through the mechanism of the Programme Local d'Habitat (PLH) which – independently of the Contrat – is a way for communes to reach agreement in a given area on priorities for housing. Since 1991, the PLH has been given a particular role in preventing concentrations of social housing in peripheral communes, in the fight against exclusion (Gaudin, 1992; Goodchild, 1993). The Contrat thus incorporates what was originally a separate mechanism in the interest of integrated policy.

In Lille, the PLH started in January 1992 and is due for completion in December 1993; it is of considerable complexity (Agence de Développement et d'Urbanisme, 1993). This is partly because the Lille conurbation's housing problems include both nineteenth-century terraces in the older centres and high-rise estates of the 1960s on the outskirts. It is also because the conurbation's division into 87 communes makes policy moves difficult. Meticulous analysis has been undertaken, and the problems identified, but so far, there is a conspicuous absence of commitment to geographical priorities for action. The key decisions on how policy will be put into effect are likely to take place in a round of horse-trading. This, coupled with the novelty of preparing a global housing strategy, will mean that at least four years will have elapsed between the signing of the Contrat and action on the ground.

Also conspicuous is the fact that for all that the PLH is part of the Contrat, discussions on housing appear to be taking place in isolation from other aspects of urban policy.

Conclusion

The Contrat d'Agglomération for the Lille conurbation thus represents an exciting attempt to co-ordinate policy in many different fields. This in itself makes it worthy of attention in other parts of Europe. Nevertheless the Contrat d'Agglomération and the CDV programme as a whole is as much part of a policy continuum as it is a radical departure. There is continuity in the content of elements of the contract; and though inter-sectoral working and global strategy are given vigorous emphasis, they form part of a long-standing concern in France to create over-arching structures to give rationality to difficult policy problems. This continuity may be an important means of ensuring the programme's success.

In the content and the procedures of the CDV there may be some general lessons to be learnt, therefore, even if the specific problems of a conurbation like Lille may vary considerably in detail from those faced in any of the City Challenge authorities in Britain.

We suggested earlier, that there were three areas of the process that merited careful attention. The first of these has to do with mechanisms. The use of the contract as the means of articulating policy is, we noted, capitalising on an instrument already in currency. This, we would argue, is not just because it is convenient to maintain continuity, or that it is a particularly good means of implementing the particular policy objectives, though both may be true. The contract appears to have a specific role in managing the fragile ecology of intercommunal working and the relationships between the many layers of local administration. This unwritten agenda for the contract may have a distorting effect on the eventual application of policy.

The second key area has to do with relationships. The relationships between actors and agencies in French local administration is extremely complex. The complexity is not merely a question of the competition between communes and the imbalance between peripheral and central areas. It has to do also with the relationship between directly elected bodies like the municipal councils and the general councils of the départements, delegated authorities like the communauté urbaine, and with local and central technical services. The exact nature of these relationships will have a direct bearing both on the content of and on the ability to implement the policy. The slow progress with implementing the Contrat d'Agglomération appears to be a result of the enormous inertia created by these relationships.

The final area of importance has to do with agencies. The Agence de Développement et d'Urbanisme has an important initiating and co-ordinating role, but its effectiveness is directly determined by the way in which the other actors regard it. Moreover it relies upon other organisations, which have their own lines of accountability, to carry out the policy. The issue of who does what is likely to be crucial to the success of the policy.

The significance of all of this goes beyond the immediate concerns of the Lille conurbation and indeed even of France itself. Mechanisms, agencies and relationships clearly have an impact both on how policy is made and on the likelihood of achieving policy objectives. We would argue that from an analysis such as this an agenda for comparison with British policy can be set.

15

CROSS-NATIONAL URBAN POLICY TRANSFER – INSIGHTS FROM THE USA

Robin Hambleton

Introduction

Cross-national urban policy transfer, particularly Anglo-American urban policy transfer, is nothing new. Thus, within months of the launch of UK urban policy in the late 1960s a conference of British and American social scientists and administrators convened at Ditchley Park, an old country house in Oxfordshire, to consider what lessons Britain could learn from the American social reform programmes of the 1960s (Marris, 1987). The conference, which influenced the shape and direction of the Community Development Project launched by the Home Secretary a few months earlier, stemmed from a meeting between Richard Nixon and Harold Wilson in 1968. The political leaders had agreed that 'the two countries should look together at some of the domestic and social problems faced by their governments'.

Turning to a more up-to-date example, it is to be expected that the UK government and opposition parties are keen to learn about the Clinton administration's policies for 'putting people first'. In the book that Bill Clinton and Al Gore published during their election year – *Putting People First. How We Can All Change America* – the outlines of an approach to the renewal of government at all levels were set out (Clinton and Gore, 1992). The President has endorsed a new book which describes what this transformation might involve in practice – *Reinventing Government: How the Entrepreneurial Spirit is Transforming the Public Sector* (Osborne and Gaebler, 1993). Only months elapsed before David Osborne was flown across the Atlantic to advise the Treasury and senior figures in UK central and local government (Arnold-Forster, 1993).

Elsewhere I have suggested that there are a number of limitations to the 'reinventing government' thesis – for example, cutting edge UK local authorities have been applying many of the ideas put forward in the book for years, the examination of 'successful' innovations is flimsy, and the political analysis of the role of government in modern society is weak (Hambleton, 1993a). Reservations of this kind are, however, unlikely to discourage enthusiasts from extolling the virtues of 'entrepreneurial government' in a fairly unthinking way.

This chapter examines the notion of cross-national urban policy transfer. It will be suggested that it is an important area for social scientific research not least because the process can provoke fresh thinking. Moreover, despite its significance, the process of cross-national policy transfer is imperfectly understood. The focus is on transatlantic policy transfer because the UK/USA dialogue in relation to urban policy and urban government continues to be extremely influential. This is not to say that urban policy exchange with other countries is unimportant. Rather, by considering transatlantic urban policy transfer in some detail, it is hoped that the chapter will generate insights which can be examined in other cross-national contexts.

Cross-National Public Policy Comparison

Why bother to compare urban policies in different countries? A review of experience suggests that there are four main reasons. First, many of the social and economic problems faced by urban authorities in different countries are similar. For example, in both the UK and the USA economic change is having an increasingly uneven impact – some regions are prospering while others are struggling. Moreover social divisions within regions and cities are growing. The alarming outbreaks of urban violence and destruction which swept through Los Angeles and a number of other cities in May 1992 drew public attention to the fact that all is not well with city living in America. As referred to in chapter 1 there have been numerous outbreaks of unrest in the disadvantaged areas of a string of UK cities in the years since the major urban rioting of 1980 and 1981. These outbreaks have always taken place in the relatively poor areas of cities where people feel powerless and frustrated. The trend towards growing social polarisation and urban unrest in both countries raises important questions about how local authorities handle rapid economic change and how they manage the social and racial tensions created by economic restructuring (Hambleton, 1990a).

Second, whilst the policy responses to these problems have often been similar, there are also striking differences – particularly in approaches to city management. These differences can, in themselves, provoke fresh thinking if the experiences and approaches of different cities and governments can be juxtaposed in a reasonably organised way. As Rose (1993) argues, the examination of approaches elsewhere can be a source of inspiration. There are, of course, substantial obstacles to systematic cross-national comparison in the field of urban policy and urban government. Major political, cultural and economic differences need to be recognised. It follows that it is highly unlikely that approaches which seem to work well in one city or country can be transplanted in a mechanistic way to another.

This is not, however, to imply that cross-national urban policy transfer is doomed. On the contrary there are many examples of urban initiatives being transferred from one country to another – for example, Urban Development Action Grants from the USA to the UK and Enterprise Zones from the UK to the USA. Moreover, general concepts and ideas, as well as specific policy instruments, have also been exchanged. In particular, we can note the way UK urban regeneration strategies in the 1980s were heavily influenced by largely American ideas about the potential of private sector, property-based renewal policies (Barnekov *et al.*, 1989; Hambleton, 1990a; Hambleton, 1990b; Parkinson, 1990).

A topic which is currently exciting a good deal of interest on both sides of the Atlantic is the idea of 'reinventing government'. Enthusiasts claim that reinventing government seeks to shift debate away from the question 'how much?' government to focus on 'what kind?' of government. On this analysis the fundamental problem is not too much government or too little government – it is that we have 'the wrong kind of government.' In other words the instrument we are using to solve collective problems and meet society's needs is outdated – hence the need for reinvention. Osborne and Gaebler (1993) suggest various principles which should guide this process of reinvention and their first one is catalytic government. This stresses that government should focus on 'steering rather than rowing' – on guiding and directing rather than concentrating on service delivery.

In the UK context many of the ideas associated with the reinventing government debate in the USA have surfaced in ongoing discussions about the changing role and function of local government (Hambleton, 1993a). Indeed some policy-makers take the view that debates about the development of an 'enabling role' for local government, which commenced in 1987/88, prefigured many of the ideas which were to emerge somewhat later in the US context. At risk of considerable over-simplification it can be suggested that UK local authorities have been shifting their role from concentration on the direct provision of a range of public services towards an orchestrating role involving outward-looking strategies designed to shape the local social and economic environment. Rather than 'working within the boundaries' of the public sector, local authorities are increasingly 'working across the boundaries' of the public, private and voluntary sectors. This has been a key theme in central government policy towards local government but it is also one which has gained support on the left in British politics (Hodge and Thompson, 1994). A recent advice document on the future management of local authorities sent to all UK council leaders and chief executives reinforces these ideas by suggesting that an authority putting the emphasis on community governance should be prepared not only to engage in the direct provision of services but also to work with a wide range of external organisations – public, private and voluntary – to meet local needs (Local Government Management Board, 1993).

The third argument in favour of cross-national public policy comparison has been identified by Wolman (1993) as policy learning. It represents a form of natural experiment in which the results of applying differing policies to similar problems can be evaluated. Policy learning does not imply direct policy transfer because there are a variety of reasons why policies which 'succeed' in one country may not work in another. In addition, comparative analysis can enhance policy learning by expanding the set of policy alternatives available for consideration:

> Comparison can not only suggest different programs, methods or techniques but, perhaps more importantly, different ways of thinking about problems and policies and different approaches.
>
> (Wolman, 1993, p. 14)

A fourth argument, also identified by Wolman (1993), is that comparative analysis also contributes to 'system understanding'. He suggests that only through contrast with other countries can one understand one's own country and what is unique about it. Thus, areas kept off the public policy agenda in one society may be extremely influential in others. Comparison also permits a better understanding of how and why units of government (local governments, nations, etc.) differ from one another. It will be suggested later that, in the urban policy context, cross-national comparison of governmental systems is just as important as cross-national study of policies and initiatives.

So much for the main reasons for embarking on cross-national urban policy comparison – how does such comparison take place? It is helpful to distinguish three levels of cross-national public policy exchange (Hambleton and Taylor, 1993). First, much of the practical exchange of ideas and experience has been between individuals and/or small groups. Contacts are made, visits are arranged, mutual learning takes place and the visitors return to their home country to make whatever use they can of what they have discovered. This model can be enormously powerful for the individuals involved. But it is often not easy for those who participate in such exchanges to translate what they have learnt into appropriate initiatives in their home country. When they return they may feel isolated, they may find colleagues are unsympathetic to the imported ideas, and they almost certainly lack advice on how to make international policy transfer work on the ground.[1]

Second, at another level, there is an increasingly active cross-national academic exchange of ideas and approaches. The field of comparative government and politics is well established and numerous texts are available which attempt to compare and contrast alternative approaches in different countries – not just to understand the variety of political systems but also to formulate and test hypotheses about the political process (Hague and Harrop, 1987; Heidenheimer, Heclo and Teich Adams, 1990). While most of this literature is restricted to comparison of national politics and government, there has in recent years been a welcome growth of interest in comparative urban policy and city government. Several of these texts focus on Anglo-American comparisons (Barnekov et al., 1989; Hambleton, 1990a; Logan and Swanstrom, 1990; Wolman and Goldsmith, 1992). These various cross-national studies are useful in providing frames of reference within which discussion of the possibilities for urban policy transfer can take place. However, a limitation is that, with some exceptions, they tend to be written at a fairly general level. As a result, it may be difficult for the busy policy-maker to see what relevance the discussions have for decision-making in the here and now.

Between these two levels there is a third, and relatively new, approach to comparative policy analysis which has the potential to make a useful contribution both to academic discourse about urban policies and urban governance and also to pressing public policy debates about alternative strategies for improving the quality of life in cities in the short term. It involves a structured exchange between policy-makers, managers, various interest groups and academics. It is a process which takes a variety of forms but it is increasingly likely to involve city-to-city dialogue rather than national government to national government exchange. New cross-national urban networks – for example, the Eurocities network in Europe – are emerging which open up new opportunities for cross-national urban policy exchange of this kind.[2]

Examining Cross-National Policy Transfer

It is possible to argue that, to date, cross-national policy transfer has relied more on subjective impressions than analysis – this certainly seems to be the case in the sphere of urban policy. Inevitably, visits to any other country to look at what seem to be promising initiatives expose policy-makers to show-case examples, enthusiasts and advocates rather than to the complex reality of policy implementation. City councils and chambers of commerce, once absorbed by the hype of 'civic boosterism', cannot be expected to draw attention to urban policy failure or neglect. When enthusiasm becomes an important element of policy implementation, dispassionate analysis is often the casualty. Thus, despite the rhetoric about US urban 'success stories', evidence about the actual impact of American urban regeneration schemes is far from comforting. Areas of astonishing urban devastation often lie only a few minutes' walk from the showpiece 'renaissance' sites (Hambleton, 1991).

Wolman (1993) has developed a three-part framework which points towards a more systematic approach in relation to cross-national urban policy transfer. It poses three critical questions:

(1) Are the problems to which the policy is to be addressed in the UK similar to those to which it was addressed in the USA? If not, are the problems to which the policy is to be applied in the UK nonetheless susceptible to the policy?

(2) To what extent was the policy 'successful' in the USA?

(3) Are there any aspects of a policy's setting in the USA which are critical to its success there, but which are not present, or are present in a different form, in the UK?

These three questions are now considered in turn. Notwithstanding the existence of substantial differences many of the 'problems' experienced by people living in cities are similar in both the UK and the USA (Hambleton and Taylor, 1993). However, great care has to be taken over problem definition for policy confusion arises if this aspect is neglected. Wolman uses local economic development to illustrate this point. He notes how local politicians and officials in the USA usually see the problem which local economic development policy addresses as primarily fiscal – the need to strengthen the local authority's tax base by increasing the amount of economic activity located within the local authority's boundaries. In contrast, political leaders and local government officers in the UK see the local economic development problem more as one of unemployment – the need to provide decent job opportunities for the area's residents. These divergent problem definitions flow from institutional, historical and policy differences. Wolman concludes:

Given the differences in problem definitions, it is quite possible that a successful local programme in the American context (for example, one that increases the local municipal tax base) may be quite unsuccessful from the British perspective (for example, if new employees commute to the area from outside and the unemployment rate of residents in the local authority is not affected).

(Wolman, 1993, p. 19)

How successful have past policies been? As numerous contributions to this volume attest this is far from easy to establish. Against what criteria should the policy be evaluated? In the context of cross-national policy transfer the task is more complex than assessing the policy on its own terms. Whether or not the policy achieves its stated intentions, whilst important, is not the primary focus of interest. Rather, the central question revolves around the extent to which the policy has successfully tackled the problem *as defined in the UK*. This may or may not be the same as the stated US policy objectives.

The research and evaluation task has, somehow, to accommodate the fact that stakeholders in a given policy may have a vested interest in showing it to be a success. As cities become increasingly concerned to project themselves in a global market-place for inward investment they, perhaps unavoidably, have a tendency to be 'economical with the truth'. Moreover, the information needed to assess policy performance may simply not be available. For example, Meyer has pointed out not only that different nations classify data in different ways, but that standard statistical sources are not up to the job – critical distinctions are often not made and the geographical units of analysis may be unhelpful (Meyer, 1993). These various obstacles to cross-national policy evaluation should not be underestimated.

Finally, there is the question of the policy setting. Some features of the policy setting may reflect aspects of the host society as a whole, while others may be specific to particular policies, or even to particular cities. Ironically, the distrust of federal government has also created local administrations which are much stronger than the UK's own *vis-à-vis* the centre. The US citizen would never tolerate the central interventions experienced in the UK. Compare the fact that there was no federal Department of Education before the Carter administration with the extraordinary centralisation of power in Whitehall which has taken place in the years since 1979. There are costs and benefits to this. In the USA there is greater institutional fragmentation and there is difficulty in spreading good practice across the country. But the greater freedom of local government, for example to raise taxes locally and to raise money through bonds, has provided resources for new initiatives and also gives business more of a stake in the local community. On the other hand, whilst US local authorities have more policy freedom, we are usefully reminded of the enormous fiscal pressure that city governments in the USA now find themselves under. It is in the USA that city governments have gone bankrupt, not the UK.

Another important difference in the policy setting is the emergence in the USA of significant intermediary bodies which can lever funds, provide support, act as a catalyst and generally provide the infrastructure on which community-based and other initiatives can be run. A well-resourced intermediary operating in the grey area between commercial and charitable objectives is invaluable in enabling local aspirations to be fulfilled at minimum risk or cost.

The policy setting is, of course, heavily shaped by the prevailing values of the host society. Important political, cultural, social, economic, racial, legal, historical and geographical differences need to be recognised. Differences in the policy setting will almost certainly mean that successful cross-national policy transfer will involve *adaptation* of policies rather than mere emulation. This is particularly the case in the sphere of local government because there is variation within each country (Hambleton, 1993b).

Having discussed the potential of cross-national urban policy analysis and hinted at some of the obstacles, the remainder of this chapter concentrates on exploring the possibilities for learning from US experience with city government. Given the fact that the UK central government has been encouraged to take new powers to enable local authorities to experiment with new forms of city management this is an area which deserves close scrutiny (Department of the Environment, 1993).

Urban Policy and Urban Governance

The review of the history of urban policy in chapter 1 suggested that central government commitment to urban policy has wavered. The Audit Commission study of urban regeneration and economic development was surely right to be critical of the proliferation of central government urban initiatives saying that 'government support programmes are seen as a patchwork quilt of complexity and idiosyncrasy' (Audit Commission, 1989, p. 1). It is widely recognised, then, that urban policy, if it is to be effective, must involve rather more than a string of central government initiatives.

This analysis gains strength from the work of US urban scholars who have argued that transforming urban governance must be a central plank of national urban policy. The following statement made about US experience applies just as well to the UK situation:

> The central element of the present urban crisis is not one of high profile problems with wide media coverage (e.g. drugs, homelessness, tax rebellions, or violent protests in the streets) but whether democratic governance itself can be maintained at the urban level. Since the late 1970s, however, deliberate and systematic policies have been adopted (by central government) to reduce or withdraw fiscal support from urban governments and to subordinate the welfare of cities to corporate profit maximisation.
>
> (Warren, Rosentraub and Weschler, 1992, pp. 401–2)

This process threatens, in both countries, to undermine the power of locally elected authorities to draw all citizens into the urban planning and governing process. In the UK context it can be claimed, with some justification, that a new magistracy – a non-elected élite – is assuming responsibility for a large part of local governance (Stewart, 1992; Weir, 1994). The growth of quangos (quasi autonomous non-governmental organisations) coupled with the multiplication of specific 'strings attached' schemes, which enable ministers to impose their priorities regardless of the wishes of local elected councillors and local people, has created a 'democratic deficit'.

Fortunately some valuable rethinking is currently taking place regarding the role, function and organisation of local authorities. It is encouraging to see that the government's consultation paper on the internal management of local authorities discusses a broad range of options and endorses the idea that experience abroad 'might merit further consideration, subject to any necessary adaptations to suit this country's traditions' (Department of the Environment, 1991, p. 11). Also encouraging is the recent report on the political management of local authorities prepared by the Department of the Environment and the local authority associations which recommends that the government should pass legislation to enable local authorities to experiment with a variety of forms of internal management (Department of the Environment, 1993). It can be suggested, therefore, that considerable opportunities for reshaping UK systems of urban governance may now be opening up – and not just in the areas affected by local government reorganisation. As part of the rethinking process it is desirable to examine experience abroad. For the reasons outlined earlier the discussion below concentrates on US experience, but approaches in Europe and elsewhere also deserve consideration.

US Models of Urban Government

Before examining the different forms of US government it is important to make three general points.[3] First, US local government is extremely fragmented. Second, the majority of American city governments use non-partisan elections, that is candidates are not identified by party on the ballot. Third, in general, the number of politicians on a city council is *much* smaller than in the UK. These striking differences reinforce the importance of adapting potentially useful ideas to the UK policy setting – importing US models 'off the shelf' is not on.

Texts on US urban government usually note that there are three basic forms: the mayor-council form, the council-manager form and the commission form (Ross, Levine and Steadman, 1991, pp. 83–104). The mayor-council form, which is the most popular system of city government in the USA, comprises a directly elected mayor and a separately elected council. Under the council-manager plan, the city council appoints a professionally trained manager, who is then given responsibility for running the daily affairs of the city. Under the commission form of government the same small group of people (usually five to seven) who comprise the city council also act as administrative heads of the various city departments.

From the point of view of current UK debates it is helpful to focus attention on variations within the first two models. The third model can be safely ignored. It has declined in popularity – in 1981 less than 3 per cent of US cities used the commission form – and it can be criticised for encouraging rampant departmentalism.

In the mayor-council form there is a separation of powers – the mayor has executive responsibility and the council exercises a legislative function. There are two broad types of mayor-council government. Figure 15.1 illustrates the strong-mayor arrangement in which the mayor has formidable administrative powers as compared to those of the council. In effect the mayor serves as the directly elected chief executive and, as a result, is highly visible. The strong mayor has the power to appoint chief officers and to veto legislation passed

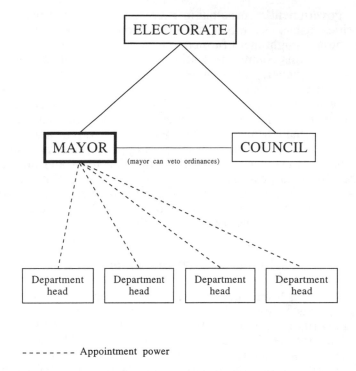

Figure 15.1   The mayor-council structure (strong mayor).

by the council. Chief officers are clear that, if the mayor loses an election, they will lose their jobs.

Figure 15.2 illustrates the weak-mayor form of government. The mayor is still directly elected but, in other respects, this arrangement resembles the existing UK system in that the formal centre of power lies in the council. In particular it is the council that appoints chief officers. There is, however, no sharp line between strong and weak mayors – rather there is a continuum.

In the council-manager form of government, as illustrated in Figure 15.3, there is no separation of powers between a political executive and a legislative body. The elected council appoints a city manager who, in turn, is directly in charge of departmental chief officers and supervises their performance. City managers have, on the whole, rather more authority than UK local authority chief executives and carry out a broad range of leadership activities. In some ways the city manager resembles the managing director of a private company.

The council-manager form can, however, create a leadership gap. City managers, because they are not elected, cannot provide political leadership. To combat this problem many council-manager cities have modified their structures along the lines shown in Figure 15.4. Here a directly elected mayor is introduced to give a political lead to the work of the city manager (Frederickson, 1989). In this context the mayor is likely to act as a facilitator rather than an executive (Svara, 1990, pp. 106–21).

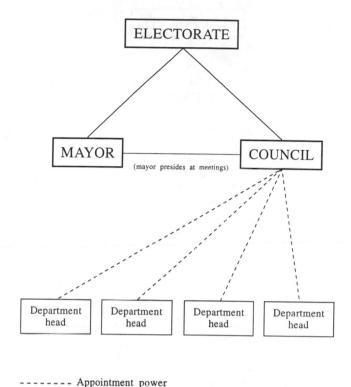

Figure 15.2 The mayor-council structure (weak mayor).

Under new powers, which it is hoped the government will provide, it should be possible for a given UK local authority to introduce an *adapted* version of any one of the four models outlined in Figures 15.1 to 15.4. The stress here is on adaptation. In particular there is a need, in the UK context, to strengthen the roles of *all* councillors and, as part of this, to develop more decentralised forms of working (Burns, Hambleton and Hoggett, 1994). This does not mean, however, that we can learn nothing from US models. As the following two examples show the USA can offer the UK new ideas, particularly in relation to city leadership and co-ordinated management.[4]

Baltimore: a Strong-Mayor City

Baltimore, which is the largest city in the State of Maryland, has a population of around 750,000. The population has dropped from 905,000 in 1970 – a decline that reflects a shift towards smaller families, the loss of manufacturing jobs and the flight to the suburbs by many middle-class residents. The city now confronts severe socio-economic problems. Whilst it attracts many of the region's highest income workers to offices in the rejuvenated downtown area near the Inner

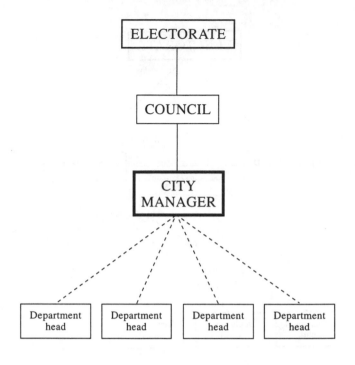

------- Appointment power

Figure 15.3 The council-manager structure.

Harbour, it cannot tax many of these people as they live in the surrounding suburban jurisdictions.

Baltimore has a strong-mayor form of government with a structure resembling Figure 15.1. The mayor, who is elected at large for a four-year term, has the power to: propose the city's budget, appoint and dismiss departmental and agency heads, and veto legislation passed by the city council (although if he or she does this the council could override such a veto with a 75 per cent majority). The council consists of 19 members - approximately one for every 40,000 people in the city – who are elected to serve councilmanic districts. In addition, the president of the city council and the financial comptroller are elected on a city-wide basis.

The Baltimore model has four main strengths. While some UK observers may feel that too much power is concentrated in the hands of the mayor, there can be no denying that 'clout gets results'. The successful regeneration of the Charles Centre/Inner Harbour area, over a thirty-year period and involving a succession of strong mayors, has established itself as a classic story of bold municipal leadership not least because the mayor can deliver. Second, the mayor is an extremely high profile public figure – it is transparently clear who is in charge. Supporters of the strong-mayor model argue that this clarifies accountability, heightens public interest in

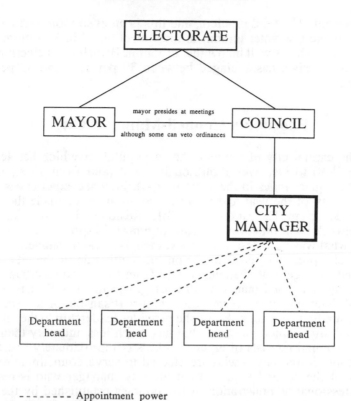

Figure 15.4   The council-manager structure with a mayor.

civic affairs and allows mayors to provide progressive role models for young people.

Third, the mayor is extremely effective in influencing the behaviour of other public and private institutions in the city through, for example, the Baltimore Development Corporation. Fourth, the mayor is able to use his office to campaign at national level on behalf of cities. This is well demonstrated by the present mayor, Kurt Schmoke, who has a national profile and is able to influence decisions well beyond the frontiers of the city.

On the downside the Baltimore model has, perhaps, three main defects. As one council member put it: 'All the city council can do is slap the mayor on the wrist – it can't make policy'. It is difficult to imagine a UK city council accepting such a diminution in its power and authority. This is not, however, a persuasive argument against importing the strong-mayor model. Rather it implies a need to *adapt* the model and give the council more powers *vis-à-vis* the mayor.

A second problem is that, while there are numerous community-based, neighbourhood initiatives, like the one in Sandtown Winchester, the city hall bureaucracy appears to be over-centralised. The structure of departments and agencies reflects historic evolution rather than the needs of modern purposeful city management. The third problem, and this is an even more serious concern in Phoenix, is the low voter turnout in local elections – in Baltimore it is less than 20 per cent. This is well below the turnout in British local elections which, whilst not impressive, has averaged between 39 per cent and 53 per cent in recent years.

Phoenix: a Council-Manager City

Phoenix, the capital city of Arizona, has a population which has leapt from 107,000 in 1950 to just over 1 million in 1992 (and from 17 square miles to a vast 428 square miles in the same period). Phoenix experiences sunshine nearly every day of the year and many corporations have made the city their national headquarters. The car dominates life. Residential densities are low and the road network takes the form of a vast, regular grid spread out across a huge expanse of what was once desert. Phoenix, then, is a classic sunbelt city. Whilst there are social problems, particularly on the south side of the city, the main planning and management challenges stem from the speed of urban growth.

Phoenix has a council-manager form of government with a mayor – see Figure 15.4. The mayor is elected on a non-partisan ballot to represent the entire city for a four-year term. He or she provides leadership to the city council and chairs all city council meetings, but has much less authority than a strong mayor. The council consists of eight members – approximately one for every 125,000 people in the city – who are elected to serve councilmanic districts. A key task of the council is to appoint the city manager who is responsible for the professional administration of the policies established by the council. The city manager has direct line management responsibility for all city staff. In addition to the officer structure the city has some 56 boards and commissions which gather public inputs into the policy-making process. Virtually all these boards are advisory.

The Phoenix approach to city government has four main strengths. The corporate processes for assessing needs, allocating resources and monitoring and managing performance are sophisticated. The budget document links expenditure and staffing to programme goals and major service levels and the management culture stresses the importance of constantly searching for improvements in quality and productivity. Second, the city makes intelligent use of competition. The city does not contract out services for the whole of the city. It retains an in-house capacity – in, for example, refuse collection – and this drives public service improvements. The success of the Public Works Department in winning contracts against the private sector is impressive (Osborne and Gaebler, 1993, pp. 76–80). Third, the city has developed a wide range of mechanisms for consulting the public – through, for example, Boards and Commissions. Such citizen panels could easily be introduced into the UK. Fourth, the city has an active policy of encouraging volunteerism – volunteers enjoy a number of benefits including training, accident insurance and reimbursement for certain expenses.

In terms of transfer to the UK context the Phoenix model has, perhaps, four main drawbacks. First, the representative democracy is fragile. There are far too few councillors – how can a city politician speak adequately for all the diverse interests and needs of 125,000 people? Moreover the level of voter turnout at 17 per cent of registered voters (which is 12 per cent of adults owing to under-registration) is disturbingly low. Second, Phoenix has social problems but does not appear to have a social policy. Third, the form of management is highly centralised. Whilst some rethinking is now taking place there is little evidence of moves to devolve budgets down the hierarchies within city departments to neighbourhood level. Fourth, some UK observers may feel that the structure concentrates too much power in the hands of the city manager. However, as with the strong-mayor model, it does not follow that the arrangement should be rejected. It would be possible, for example, in the UK context to create scrutiny committees of councillors to monitor the work of the city manager and his or her staff.

Conclusion

In this chapter it has been suggested that there has been a good deal of policy exchange between the USA and the UK in the field of urban policy and government. In recent years, while the national urban policies in both countries continue to attract attention (Hambleton, 1994b), a re-examination of the institutions of urban governance has commenced and this is to be welcomed. Norton Long, a leading American political scientist, has argued persuasively that urban policy must strengthen the links between citizens and their cities:

> Governments are territorial, and the future of this nation's cities depends on there being enough of their inhabitants with a strong commitment to ensuring they have a future.

> (Long, 1993)

In line with Long's analysis this chapter has argued that effective strategies for tackling urban problems will require not just imaginative national urban policies but also a transformation of the institutions of urban governance. Revitalising local democracy needs to be seen as an integral part of any new strategy for cities. The discussions now taking place under the banner of reinventing local government and/or developing the orchestrating role of local government could well turn out to have a more lasting significance than the latest urban policy initiatives emanating from Washington DC or Whitehall.

Given that the UK is in the process of developing new forms of urban governance, not least as part of local government reorganisation, it is well worth while studying experience abroad. The chapter has suggested that American cities tend to be more prepared to innovate and create new institutional forms than their UK counterparts. As UK city councils rethink their approaches to city leadership, community representation and urban development they could do worse than examine the experience of leading US cities. The diversity of models is refreshing and contrasts sharply with the relative uniformity of approaches to city government in the UK. The brief commentaries on Baltimore and Phoenix suggest that, whilst there are pitfalls to avoid, there are exciting possibilities to explore.

A key theme of the chapter is that cross-national comparison is likely to be fruitful. This is not because such comparisons are likely to provide ready-made answers. On the contrary a successful approach will involve more systematic evaluation than has tended to be the case in the past. Some suggestions on how to improve the process of cross-national policy transfer have been outlined. In closing it should be noted that it is likely to be more difficult to transfer institutional models as compared to policies and initiatives. Forms of urban government are more deeply rooted in society than policy. Substantial adaptation is likely to be necessary to an imported governmental form if it is to be tuned to local values, culture and politics. This challenging task is now very much on the public policy agenda in the UK.

Notes

1. This conclusion derives from discussions with a large number of UK *Harkness Fellows* (who spend six to nine months in a US public policy role and then return to the UK) and the author's involvement with a US Study Tour of four US cities organised by the UK *Local Government Management Network* in June 1993.
2. The conference *People in Cities: a Transatlantic Policy Exchange* organised at the School for Advanced Urban Studies, University of Bristol in October 1991 provides another example of this model. It involved political leaders from US and UK cities, civil servants, officers from local government and quangos, Harkness Fellows and representatives from voluntary organisations as well as academics. Many of the papers presented are included in Hambleton and Taylor (1993).
3. These points are discussed in more detail in Hambleton (1993b) and Stoker and Wolman (1992).
4. These outlines are based on visits made to Baltimore and Phoenix in Spring 1993 and I would like to acknowledge the very helpful assistance provided by politicians, officers and others in both cities. A more detailed appraisal is to be found in Hambleton (1994a).

REFERENCES

Agence de Développement et d'Urbanisme (1993) *Pour une Agglomération Solidaire*, Lille.

Anderson, T. (1992) *The Territorial Imperative*, Cambridge University Press.

Archer, C. (1994) *Organizing Europe: The Institutions of Integration*, London, Edward Arnold.

Arnold-Forster, J. (1993) Gore's own guru, *Local Government Chronicle*, 4 June, pp. 14–15.

Arrow, K. J. (1951) *Social Choice and Individual Values*, New York, NY: Wiley.

Association of Metropolitan Authorities (1993) *Local Authorities and Community Development: a Strategic Opportunity for the 1990s*, London: Association of Metropolitan Authorities.

Atkinson, R. and Moon, G. (1994) *Urban Policy in Britain: the City, the State and the market*, Basingstoke: Macmillan.

Audit Commission (1989), *Urban Regeneration and Economic Development: the Local Government Dimension*, London: HMSO.

Audit Commission (1991) *A Rough Guide to Europe: Local Authorities and the EC*, London: HMSO.

Audit Commission (1993) *Passing the Bucks: the Impact of Standard Spending Assessments on Economy, Efficiency and Effectiveness*, London: HMSO.

Bach, J. (1992) Policy evaluation, *Regional Studies*, Vol. 25, no. 3, pp. 262–7.

Bailey, N. (1994) Towards a research agenda for public-private partnerships in the 1990s, *Local Economy*, Vol. 8, no. 4.

Ballard, R. and Kalra, V. S. (1994) *The Ethnic Dimensions of the 1991 Census: a Preliminary Report*, Manchester Census Group, University of Manchester.

Barnekov, T. *et al.* (1989) *Privatism and Urban Policy in Britain and the United States*, Oxford University Press.

Barnekov, T. *et al.* (1990) *US Experience in Evaluating Urban Regeneration*, London: HMSO.

Barrett, S. and Fudge, C. (eds.) (1981) *Policy and Action*, London: Methuen.

Barrett, S. M., Stewart, M. and Underwood, J. (1978) *The Land Market and the Development Process*, SAUS Occasional Paper No. 2, School for Advanced Urban Studies, Bristol.

Bartholomew, D. J. (1988) *Measuring Social Disadvantage and Additional Educational Needs*, London, LSE Department of Statistical and Mathematical Sciences, a report to the Department of the Environment.

Beatley, T. (1994) *Ethical Land Use*, Baltimore: Johns Hopkins University Press.

Begg, I., Moore. B. and Rhodes, J. (1986) Economic and social change in urban Britain and the inner cities, in V. Hausner (ed.) *Critical Issues in Urban Economic Development*, Vol. 1, Oxford: Clarendon Press.

Bell, D. (1990) *Data Sources for Area Prioritisation: Section A – Review and Analytical Topics*, Edinburgh: DG Information Services.

Bendick, M. and Egan, M.L. (1993) Linking business development and community development in inner cities, *Journal of Planning Literature*, Vol. 8, no. 1, pp. 3–19.

Bennett, R., Wicks, P. and McCoshan, A. (1994) *Local Empowerment and Business Services: Britain's Experiment with Training and Enterprise Councils*, London: UCL Press.

Betts, C. (1992) Plaid revolution hopes, *Western Mail*, 15 August.

Beynon, H. (1988) Regulating research: politics and decision making in industrial organisations, in A. Bryman (ed.) *Doing Research in Organisations*, London: Routledge, pp. 21–33.

Bidwell, C. (1993) *Maastricht and the UK*, Public Affairs Consultants Europe Limited.

Blunkett, D. and Jackson, K. (1987) *Democracy in Crisis: the Town Hall Responds*, London: Hogarth Press.

Boddy, M. (1992) Positive action employment and training: developing local initiative, *Local Economy*, Vol. 7, no. 1, pp. 51–63.

Boddy, M., Bridge, G., Burton, P. and Gordon, D. (1995) *Socio-Demographic Change and the Inner City*, London: HMSO.

Boddy, M., Lovering, J. and Bassett, K. (1986) *Sunbelt City? A Study of Economic Change in Britain's M4 Growth Corridor*, Oxford: Clarendon Press.

Booth, P. and Green, H. (1993) *Urban Policy in England and Wales and France: a Comparative Assessment of Recent Policy Initiatives*, TRP113, Department of Town and Regional Planning, University of Sheffield.

Bovaird, A., Gregory, D. G. and Martin, S. J. (1991) Improved performance management in local economic development: a warm embrace or an artful sidestep? *Public Administration*, Vol. 69, no. 1, pp. 103–19.

Boyle, M. R. (1989) Grading the state business climate report cards, *The Economic and Demographic Trends Newsletter*, Vol. 2, no. 1, pp. 1–6.

Bradford, M. G., Robson, B. T. and Tye, R. (1993) *Constructing the 1991 Urban Deprivation Index*, SPA Working Paper 24, School of Geography, University of Manchester (forthcoming 1995 in *Environmental Planning, A*).

Breheny, M. and Hall, P. (1984) The strange death of strategic planning and the victory of the know-nothing school, *Built Environment*, Vol. 10, no. 2, pp. 95–9.

Bresnan, M. (1988) Insights on site: research into construction project organisations, in A. Bryman (ed.) *Doing Research in Organisations*, London: Routledge, pp. 34–52.

Brindley, T., Rydin, Y. and Stoker, G. (1989) *Remaking Planning*, London: Unwin Hyman.

Bristol City Council (1988) *Poverty in Bristol: an update*, Bristol: BCC.

Bromley, D. W. (1991) *Environment and Economy: Property Rights and Public Policy*, p. 161.

Bryman, A. (ed.) (1988) *Doing Research in Organisations*, London: Routledge.

Bryson, J. and Crosby, B. (1989) The design and use of strategic planning arenas, *Planning Outlook*, Vol. 32, no. 1, pp. 5–13.

Burns, D., Hambleton, R. and Hoggett, P. (1994) *The Politics of Decentralisation: Revitalising Local Democracy*, London: Macmillan.

Burton, P. (1991) Challenging cities: a new approach to urban regeneration? Paper presented to European Workshop on the Improvement of the Built Environment and Social Integration in Cities, Berlin, 9–11 October.

Burton, P. and O'Toole, M. (1993) Urban Development Corporations: post-Fordism in action or Fordism in retrenchment? in R. Imrie and H. Thomas (eds.) *British Urban Policy and the Urban Development Corporations*, London: Paul Chapman.

Burton, P. and Stewart, M. (1995) Concentration and segregation: the implications for urban policy, in P. Ratcliffe (ed.) *Geographical Spread, Spatial Concentration and Internal Migration*, London: OPCS.

Cabinet Office (1994a) *Open Government: Code of Practice on Access to Government Information*, London: HMSO.

Cabinet Office (1994b) *Open Government: Code of Practice on Access to Government Information: Guidance on interpretation*, London: HMSO.

Campbell, B. (1993) *Goliath: Britain's Dangerous Places*, London: Methuen.

Campbell, M. (1990) (ed) *Local Economic Policy*, London: Cassell.

Cardiff City Council (1988) *Cardiff Bay Regeneration Strategy: the City Council's Response*, Cardiff: Cardiff City Council.

Carley, M. (1980) *Rational Techniques in Policy Analysis*, London: Heinemann Educational.

Carley, M. (1981) *Social Measurement and Social Indicators*, London: George Allen and Unwin.

Castlemilk Partnership (1994) *Mid-Term Review of Housing Strategy*, Glasgow: Castlemilk Partnership.

CEC (1987) *Urban Problems and Regional Policy in the European Community*, Luxembourg: CEC.

CEC (1990) *Green Paper on the Urban Environment*, Brussels: CEC.

CEC (1991) *Europe 2000 – Outlook for the Development of the Community's Territory*, Brussels: CEC.

CEC (1992a) *Urbanisation and the Functions of Cities in the European Community*, Centre for Urban Studies, University of Liverpool.

CEC (1992b) *Towards a Europe of Solidarity: Intensifying the Fight against Social Exclusion, Fostering Integration*, Brussels: CEC.

CEC (1992c) *Treaty on European Union*, Luxembourg: CEC.

CEC (1992d) *Towards a Europe of Solidarity*, Communication from the Commission, COM (92) 542 Final, Brussels: CEC.

CEC (1992e) *From Single Market to European Union*, Luxembourg: CEC.

CEC (1992f) *Reform of the Structural Funds: A Tool to Promote Economic and Social Exclusion*, Brussels: CEC.

CEC (1993a) *Community Structural Funds 1994–99: Regulations and Commentary*, Luxembourg: CEC.

CEC (1993b) *Green Paper: European Social Policy Options for the Union*, Brussels: CEC.

CEC (1993c) *White Paper: Growth Competitiveness and Employment*, Brussels: CEC.

CEC (1993d) *Medium-Term Action Programme to Combat Exclusion and Promote Solidarity: A New Programme to Support and Stimulate Innovation (1994–1999)*, COM (93) 435 Final, Brussels: CEC.

CEC (1993e) *Social Exclusion – Poverty and Other Social Problems in the European Community. Background Report*, Jean Monnet House, London: CEC.

CEC (1994a) *The Future of Community Initiatives under the Structural Funds*, Com (94) 46 Final, Brussels: CEC.

CEC (1994b) *Community Initiatives Concerning Urban Areas (URBAN)*, COM (94) 61 Final\2, Brussels: CEC.

CEC (1994c) *Fourth Annual Report: the implementation of the reform of the Structural Funds 1992.* Luxembourg.

Centre for Local Economic Strategies (1991) Local economic strategies, *Local Work*, no. 25, April.

CES (1985) Deprived areas beyond the pale, *Town and Country Planning*, February, pp. 54–5.

Champion, A. (1992) Urban and regional demographic trends in the developed world, *Urban Studies*, Vol. 29, nos. 3/4, pp. 461–82.

Charlton, M. E., Openshaw, S. and Wymer, C. (1985) Some new classifications of Census enumeration districts in Britain: a poor man's ACORN, *Journal of Economic and Social Measurement*, Vol. 13, pp. 69–96.

Cheshire, P. (1987) Urban policy: art not science? in B. Robson (ed.) *Managing the City: the Aims and Impacts of Urban Policy*, New Jersey: Barnes and Noble.

Cheshire, P. (1990) Explaining the recent performance of the European Community's major urban regions, *Urban Studies*, Vol. 27, no. 3, pp. 311–33.

Cheshire. P., Carbonaro, G. and Hay, D. (1986) Problems of urban decline and growth in EEC countries: or measuring degrees of elephantness, *Urban Studies*, Vol. 23, no. 2, pp. 131–49.

Cheshire, P. C. and Hay, D. G. (1989) *Urban Problems in Western Europe: an Economic Analysis*, London: Unwin Hyman.

Clapham, D. and Kintrea, K. (1992) *Housing Co-operatives in Britain, Achievements and Prospects.*

Clarke, M. and Stewart, J. (1994) The local authority and the new community governance, *Regional Studies*, Vol. 28, no. 2, pp. 201–7.

Clinton, B. and Gore, A. (1992) *Putting People First. How We Can All Change America*, New York: Random House.

Cochrane, A. (1991) The changing state of local government: restructuring for the 1990s, *Public Administration*, Vol. 69, no. 3, pp. 281–302.

Cochrane, A. (1993) *Whatever Happened to Local Government*, Milton Keynes: Open University Press.

Cole, S. (1993) Cultural accounting in small economies, *Regional Studies*, Vol. 27, no. 2, pp. 121–36.

Colenutt, B. and Tansley, S. (1990) *Inner City Regeneration: a Local Authority Perspective.* First year report on the Urban Development Corporations, Manchester: Centre for Local Economic Strategies.

Commissariat Général du Plan (1992) *Villes, Démocratie, Solidarité: Le Pari d'une Politique*, Le Moniteur/Documentation Française, Paris.

Commission of the European Communities (1991) *Annual Report on the Implementation of the Reform of the Structural Funds 1989*, Luxembourg: Office for Official Publications of the European Communities.

Communauté Urbaine de Lille (1992a) *Contrat-Cadre D'Agglomération, Le Contrat et Les Tableaux Financiers*, Lille.

Communauté Urbaine de Lille (1992b) *Contrat-Cadre D'Agglomération, Annexe Le Détail Des Actions*, Lille.

Cooke, P. and Morgan, K. (1993) The network paradigm: new departures in corporate and regional development, *Environment and Planning D*, Vol. 11, pp. 543–64.

Coombes, M. G. and Raybould, S. (1989) Developing a Local Enterprise Activity Potential (LEAP) index, *Built Environment*, Vol. 14, pp. 107–17.

Coombes, M. G., Raybould, S. and Wong, C. (1992) *Developing Indicators to Assess the Potential for Urban Regeneration*, London: HMSO.

Corporation for Enterprise Development (1991) *The 1991 Development Report Card for the States: a Tool for Public and Private Sector Decision Makers*, Washington: The Corporation for Enterprise Development.

Coulson, A. (1988) The evaluator: inquisitor, comrade or spy? *Local Economy*, Vol. 2, no. 4, pp. 229–36.

Coulson, A. (1990) Evaluating local economic policy, in M. Campbell (ed.) *Local Economic Policy*, London: Cassell, pp.174–94.

CURDS (1979) *The Mobilisation of Indigenous Potential in the UK*, University of Newcastle-upon-Tyne, CURDS.

Cynon Valley Borough Council Labour Group (1994) *Cynon Valley Urban Joint Venture: a Model for Regeneration*, Evidence to the City 2020 Inquiry.

Davidson, R. (1976) Social deprivation: an analysis of intercensal change, *Transactions of the IBG*, Vol. 1, pp. 108–17.

Davies, H. W. E. (1980) *International Transfer and the Inner City*, Occasional Paper OP5, School of Planning Studies, University of Reading.

Davies, T. and Mason, C. (1986) *Shutting Out the Inner City Worker*, SAUS Occasional Paper 23, Bristol: School for Advanced Urban Studies.

Davoudi, S. (1993) Women and part-time employment training, *Local Economy*, Vol. 8, no. 1, pp. 33–42.

Davoudi, S. and Healey, P. (1994a) City Challenge: sustainable process or temporary gesture, *Environment and Planning C: Government and Policy* (forthcoming).

Davoudi, S. and Healey, P. (1994) *Perceptions of City Challenge Policy Processes*, Working Paper, Department of Town and Country Planning, University of Newcastle-upon-Tyne.

Deakin, N. and Edwards, J. (1993) *The Enterprise Culture and Inner City*, London: Routledge.

Deas, I. A. and Harrison, E. K. (1994) Hopes left to crumble, *Local Government Chronicle*, 13 March, p. 13.

Deas, I. A. and Robinson, F. (1994) *Squeezing the Inner Cities: Responses to Changing Urban Policy Priorities in Tyne and Wear*. SPA Working Paper 25, School of Geography, University of Manchester.

De Groot, L. (1992) City Challenge: competing in the urban regeneration game, *Local Economy*, Vol. 7, no. 3, pp. 196–209.

Dendrinos, D. (1992) *The Dynamics of Cities: Ecological Determinism, Dualism and Chaos*, New York: Routledge.

De Neufville, J. (1975) *Social Indicators and Public Policy*, Amsterdam: Elsevier.

Department of the Environment (1977) *Policy for the Inner Cities*, Cmnd. 6845, London: HMSO.

Department of the Environment (1987a) *An Evaluation of the Enterprise Zone Experiment*, London: HMSO.

Department of the Environment (1987b) *An Evaluation of Derelict Land Grant Schemes*, London: HMSO.

Department of the Environment (1988) *An Evaluation of the Urban Development Grant Programme*, London: HMSO.

Department of the Environment (1991a) *City Challenge Model Action Plan: Notes for Guidance on the Completion of Pacemaker Action Plans*, London: Victor

244 REFERENCES

Hausner Associates.
Department of the Environment (1991b) *Annual Report 1991. The Government's Expenditure Plans 1991–92 to 1993–94* (Cm. 1508), London: HMSO.
Department of the Environment (1991c) *City Challenge: Government Guidance* (May) London: DoE.
Department of the Environment (1991d) *Local Government Review. The Internal Management of Local Authorities in England. A Consultation Paper*, London: Department of the Environment.
Department of the Environment (1991e) *City Challenge*, Action for Cities News Release, 23 May.
Department of the Environment (1992a) *City Challenge: Bidding Guidance 1993/94*, London: DoE.
Department of the Environment (1992b) *Annual Report 1992. The Government's Expenditure Plans 1992–93 to 1994–95* (Cm. 1908), London: HMSO.
Department of the Environment (1992) *City Challenge II: Michael Heseltine looks for 20 winners*, Press Release, 18 February.
Department of the Environment (1992) *Michael Howard announces 20 City Challenge winners and four new Task Forces*, Press Release, 16 July.
Department of the Environment (1993a) *Community Leadership and Representation: Unlocking the Potential*, London: HMSO.
Department of the Environment (1993b) *Annual Report 1993: The Government's Expenditure Plans 1993–94 to 1995–96* (Cm. 2207), London: HMSO.
Department of the Environment (1993c) *Single Regeneration Budget: Note on Principles and Factsheets*, London: DoE.
Department of the Environment (1994a) *Annual Report 1994. The Government's Expenditure Plans 1994–95 to 1996–97* (Cm. 2507), London: HMSO.
Department of the Environment (1994b) *Bidding Guidance: A guide to funding under the Single Regeneration Budget*, London: DoE.
Department of the Environment (1994c) *Assessing the Impact of Urban Policy*, London, HMSO.
Department of Trade and Industry (1991) *An Evaluation of the Government's Inner Cities Task Force Initiative*, London: HMSO.
Donnison, D. (1991) *A Radical Agenda*, London: Rivers Oram Press.
Drysek, J. (1990) *Discursive Democracy: Politics, Policy and Political Science*, Cambridge University Press.
Duguid, G. and Grant, R. (1983) *Areas of Special Need in Scotland*, Edinburgh: Scottish Office Central Research Unit.
Duvall, R. and Shamir, M. (1981) Indicators from errors: cross-national, time-serial measures of the repressive disposition of governments, in C. L. Taylor (ed.) *Indicator Systems for Political, Economic and Social Analysis*, Cambridge, Mass: Gunn and Hain.
Edelman, M. (1977) *Political Language: Words that Succeed and Policies that Fail*, Institute for Research on Poverty: New York Academic Press.
Edwards, J. and Batley, R. (1978) *The Politics of Positive Discrimination*, London: Tavistock.
Eisenschitz, A. and Gough, J. (1993) *The Politics of Local Economic Policy*, London: Macmillan.
Eisenstadt, S. N. and Lamarchand, R. (eds.) (1981) *Political Clientelism, Patronage and Development*, London: Sage.
Ermisch, J. (1990) *Fewer Babies, Longer Lives: Policy Implications of Current Demographic Trends*, York: Joseph Rowntree Foundation.

European Commission (1991) *Ex-Ante Evaluation of Community Support Programmes and Dependent Programmes for the Objective 2 Areas for South Wales and Bremen*, Final Report, Brussels.

European Council (1993) *Presidency Conclusions*, Brussels.

European Foundation for the Improvement of Living and Working Conditions (1986) *Living Conditions in Urban Areas*, Dublin: EFILWC.

European Foundation for the Improvement of Living and Working Conditions (1989) *Social Change and Local Action: Coping with Disadvantage in Urban Areas*, Dublin: EFILWC.

European Foundation for the Improvement of Living and Working Conditions (1990) *Mobility and Social Cohesion in the European Community – A Forward Look*, Dublin: EFILWC.

European Foundation for the Improvement of Living and Working Conditions (1994) *Bridging The Gulf*, Dublin: EFILWC.

Fielding, A. and Halford, S. (1990) *Patterns and Processes of Urban Change in the United Kingdom*, London: HMSO.

Findlay, A., Morris, A. and Rogerson, R. (1988) Where to live in Britain in 1988: quality of life in British cities, *Cities*, Vol. 5, no. 3, pp. 268–76.

Fischer, F. (1990) *Technology and the Politics of Expertise*, London: Sage.

Foley, P. (1992) Local economic policy and job creation: a review of evaluation studies, *Urban Studies*, Vol. 29, no. 3/4, pp. 557–98.

Forester, J. (1989) *Planning in the Face of Power*, Berkley: California UP.

Forester, J. (1993) *Critical Theory, Public Policy and Planning Practice*, Albany: Suny Press.

Fothergill, S. (1988) Researching the inner city, *Local Economy*, Vol. 3, no. 2, pp. 85–95.

Frederickson, H. G. (ed.) (1989) *Ideal and Practice in Council-Manager Government*, Washington DC: International City Management Association.

Friedmann, J. (1987) *Planning in the Public Domain*, New Jersey: Princeton UP.

Friedmann, J. (1992) *Empowerment: the Politics of Alternative Development*, Oxford: Blackwell.

Friend, J. and Hickling, A. (1977) *Planning Under Pressure*, Oxford: Pergamon.

Gamble, A. (1988) *The Free Economy and the Strong State*, London: Macmillan.

Gans, H. (1993) From underclass to undercaste: some observations about the future of the postindustrial economy and its major victims, *International Journal of Urban and Regional Research*, Vol. 17, no. 3, pp. 327–35.

Gaudin, J.-P. (1992) La loi d'orientation pour la ville et la conduite des politiques publiques, *Cahiers français*, Vol. 256, May–June, pp. 51–9.

Goodchild, B. (1993) Local housing strategies in France, *Planning Practice and Research*, Vol. 8, no. 1, pp. 4–7.

Gould, S. J. (1981) *The Mismeasure of Man*, New York: W. W. Norton.

Government Offices for the Regions (1994) Bidding Guidance: a Guide to Funding from the Single Regeneration Budget, London: HMSO.

Gramlich, E. M. (1981) *Benefit-Cost Analysis of Government Programmes*, Englewood Cliffs, NJ: Prentice-Hall.

Green, A. (1994) The geography of poverty and wealth, 1981–1991, *Rowntree Findings*, Social Policy Research 55, York: Joseph Rowntree Foundation.

Green, A. E. and Champion, A. (1991) Booming towns studies: methodological issues, *Environment and Planning A*, Vol. 23, pp. 1393–1408.

Green, D. H. and Paris, D. (1992) Les PME et les politiques de développement local et régional. Etude comparée entre des régions du Royaume-Uni et de la France: le Yorkshire-Humberside et le Nord-Pas-De-Calais, in D. H. Green and K. Hayton (eds.) *Getting People into Jobs: Good Practice in Urban Regeneration*, London: HMSO.

Gregory, D. G. and Martin, S. J. (1988) Issues in the evaluation of inner city programmes, *Local Economy*, Vol. 2, no. 4, pp. 237–49.

Gregory, R., Keeney, R. and von Winterfeldt, D. (1992) Adapting the environmental impact statement process to inform decisionmakers, *Journal of Policy Analysis and Measurement*, Vol. 11, no. 1, pp 58–75.

Guichard, O. (1993) Une compétence de l'etat, *le Figaro*, 18 August.

Gurr, T. R. (1981) A conceptual system of political indicators, in C. L. Taylor (ed.) *Indicator Systems for Political, Economic and Social Analysis*, Cambridge, Mass: Gunn and Hain.

Gyford, J. (1985) *The Politics of Local Socialism*, London: George Allen and Unwin.

Hague, R. and Harrop, M. (1987) *Comparative Government and Politics*, London: Macmillan.

Hall, P. (1988) *Cities of Tomorrow*, Oxford: Blackwell.

Hall, P. (1990) Transitional stress and the crisis of those without hope, *Town and Country Planning*, December, pp. 331–3

Hall, P. (1994) Forces shaping urban Europe, *Urban Studies*, Vol. 30, no. 2, pp. 883–98.

Ham, C. and Hill, M. (1993) *The Policy Process in the Modern Capitalist State* (2nd edn) London: Harvester Wheatsheaf.

Hambleton, R (1978) *Policy Planning and Local Government*, London: Hutchinson.

Hambleton, R (1981) Implementing inner city policy: reflections from experience, *Policy and Politics*, Vol. 9, no. 1, pp. 51–71.

Hambleton, R. (1988) Policy planning reconsidered, *Journal of the Royal Town Planning Institute*, Vol. 74, no. 11, pp. 12–15.

Hambleton, R. (1990a) *Urban Government in the 1990s: Lessons from the USA*, School for Advanced Urban Studies, University of Bristol.

Hambleton, R. (1990b) Privatising urban regeneration, *Town and Country Planning*, November.

Hambleton, R. (1991a) Another chance for cities? Issues for urban policy in the 1990s, *Papers in Planning Research*, No. 126, Department of City and Regional Planning, University of Wales, Cardiff.

Hambleton, R. (1991b) American dreams, urban realities, *The Planner*, 28 June, pp. 6–9.

Hambleton, R. (1993a) Issues for urban policy in the 1990s, *Town Planning Review*, Vol. 64, no. 3, pp. 313–23.

Hambleton, R. (1993b) Reinventing local government, *Parliamentary Brief*, Vol. 2, no. 1, July, pp. 20–1.

Hambleton, R. (1993c) Reflections on urban government in the USA, *Policy and Politics*, Vol. 21, no. 4, October, pp. 245–57.

Hambleton, R. (1994a) Urban Management in US cities: Lessons from Baltimore and Phoenix, *Papers in Planning Research* No. 150, University of Wales, Cardiff.

Hambleton, R. (1994b) Lessons from America, *Planning Week*, 16 June, pp. 16–17.

Hambleton, R. (1994c) *Evaluating Neighbourhood Democracy: Reflections on a Research Project*. Paper presented at the ESRC Urban Evaluation seminar, University of Wales, Cardiff, September.

Hambleton, R. and Taylor, M. (eds.) (1993) *People in Cities: a Transatlantic Policy Exchange*, School for Advanced Urban Studies, University of Bristol.

Hansard (1994) House of Commons 31 March and 25 April, Vols. 240–2.

Harrison, P. (1983) *Inside the Inner City: Life Under the Cutting Edge*, Harmondsworth: Penguin.

Harvey, D. (1989) From managerialism to entrepreneurialism: the transformation of urban governance in late capitalism, *Geografiska Annaler*, Vol. 71B, no. 1, pp. 3–17.

Hart, D. (1991) US urban policy evaluation in the 1980s – lessons from practice, *Regional Studies*, Vol. 25, no. 3, pp. 255–61.

Hayton, K. (1990) *Getting People into Jobs: Good Practice in Urban Regeneration*, London: HMSO.

Healey, P. (1990) Policy processes in land use planning, *Policy and Politics*, Vol. 18, no. 1, pp. 91–103.

Healey, P. (1991) Urban regeneration and the development industry, *Regional Studies*, Vol. 25, no. 2, pp. 97–109.

Healey, P. (1992) Planning through debate, *Town Planning Review*, Vol. 63, no. 2, pp. 143–63.

Healey, P. (1994) Urban policy and property development, *Environment and Planning A* (forthcoming).

Healey, P. and Davoudi, S. (1993) *The Development Industry in Tyne and Wear*, Working paper no. 21, Department of Town and Country Planning, University of Newcastle.

Healey, P. and Gilroy, R. (1990) Towards a people-sensitive *Planning Practice and Research*, Vol. 5, no. 2, pp. 21–9.

Healey, P., McNamara, P., Elson, M. and Doak, J. (1988) *Land Use Planning and the Mediation of Urban Change*, Cambridge University Press.

Healey, P., Blackman, T., Cameron, S. J., Davoudi, S., Gilroy, R., Hughes, R., Marvin, S. and Woods, R. (1992) *Newcastle's West End: Monitoring the City Challenge Initiative: A Baseline Report*, Department of Town and Country Planning, University of Newcastle.

Healey, P. and Barrett, S. M. (1990) Structure and Agency in Land and Property Development Processes, *Urban Studies*, Vol. 27, no. 1, pp. 89–103.

Healey, P., Davoudi, S., O'Toole, M., Tavsanoglu, S. and Usher, D. (1992) *Rebuilding the City: Property-Led Urban Regeneration*, London: E. and F. N. Spon.

Healey, P., Cameron, S., Davoudi, S., Graham, S. and Madani-Pour, A. (1995) *Managing Cities*, London: John Wiley.

Heidenheimer, A. J., Heclo, H. and Teich Adams, C. (1990) *Comparative Public Policy* (3rd edn) New York: St. Martin's Press.

Her Majesty's Government (1977) *Policy for the Inner Cities*, London: HMSO.

Her Majesty's Government (1988) *Action for Cities*, March, London: HMSO.

Heseltine, M. (31 July 1993) First round decision letter (to Councillor Robertson, Bristol City Council).

Hillier, J. (1993) Discursive democracy in action. Paper to the 7th AESOP Congress, Lodz, Poland, July.

HM Treasury (1988) *Evaluation: a Policy Guide for Managers*, London: HMSO.

Hodge, M. and Thompson, W. (1994) *Beyond the Town Hall – Reinventing Local Government*, Fabian Pamphlet 561, London: Fabian Society.

Hoggett, P. (1991) A new management in the public sector? *Policy and Politics*, Vol. 19, no. 4, pp. 243–56.

Horton, C. and Smith, D. (1988) *Evaluating Police Work*, London: Policy Studies Institute.

Houlder, V. (1992) Docklands gets that sinking feeling, *Financial Times*, 8 May, p. 12.

House of Commons Employment Committee (1988) *The Employment Effects of Urban Development Corporations*, London: HMSO.

Hunt, D. (1993a) Speech to the CBI/WDA Conference on Urban Development in the 1990s, 25 January, London.

Hunt, D. (1993b) *The Valleys Initiative*, Hansard, Cols. 513–16, 1 April.

IFO (1990) *An Empirical Assessment of Factors Shaping Regional Competitiveness in Problem Regions*, Brussels: Institute for Economic Research, Commission of the European Communities.

Imrie, R. and Thomas, H. (1992) The wrong side of the tracks: a case study of local economic regeneration in Britain, *Policy and Politics*, Vol. 20, no. 3, pp. 213–26.

Imrie, R. and Thomas, H. (1993a) The limits of property-led regeneration, *Environment and Planning C: Government and Policy*, Vol. 11, pp. 87–103.

Imrie, R. and Thomas, H. (eds.) (1993b) *British Urban Policy and the Urban Development Corporations*, London: Paul Chapman.

Imrie, R. and Thomas, H. (1994) The new partnership: the local state and the property development industry, in R. Ball and A. Pratt (eds.) *Industrial Property: Policy and Economic Development*, London: Routledge.

Imrie, R., Thomas, H. and Marshall, T. (1995) Business organisations, local dependence, and the politics of urban renewal in Britain, *Urban Studies*, Vol. 32, no. 1.

Investment Property Databank (1993) *The IPD Property Investors Digest 1993*, IPD, London.

Jacobs, B. D. (1992) *Fractured Cities: Capitalism, Community and Empowerment in Britain and America*, New York, NY: Routledge.

Jacquier, C. (1991) *Voyage dans Dix Quartiers Européens en Crise*, Paris: L'Harmattan.

Jessop, B. (1993) From the Keynesian Welfare to the Schumpeterian Workfare State, in R. Burrows and B. Loader (eds.) *Towards a Post-Fordist Welfare State*, London: Routledge.

Johnson, D. M., Martin, S. J. and Pearce, G. R. A. (1992) *The Strategic Approach to Derelict Land Reclamation*, London: HMSO.

Judd, D. and Parkinson, M. (eds.) (1990) *Leadership and Urban Regeneration: Cities in North America and Europe*, London: Sage.

Keating, M. (1991) *Comparative Urban Politics*, Aldershot: Edward Elgar.

Labour Party (1993) *A Positive Future for the South Wales Valleys*, Labour Party Wales, Cardiff.

Lansley, S., Goss, S. and Wolmar, C. (1989) *Councils in Conflict*, London: Macmillan.

Lawless, P. (1988) *Britain's Inner Cities*, London: Paul Chapman.

Lawless, P. (1991) Urban Policy in the Thatcher decade: English inner-city policy 1979–80, *Environment and Planning c: Government and Policy*, Vol. 9, pp. 15–30.

Lawless, P. (1992) Social integration and new urban activities, *The Improvement of the Built Environment and Social Integration in Cities*, European Foundation for the improvement of Living and Working Conditions, Dublin.

Leach, S. (ed.) (1992) *Strengthening Local Government in the 1990s*.

Lee, R. (1993) *Doing Research on Sensitive Topics*, London: Sage.

Lever, W. F. (1993) Competition within the European urban system, *Urban Studies*, Vol. 30, no. 6, pp. 935–48.

Lewis, N. (1992) *Inner City Regeneration: The Demise of Regional and Local Government*, Buckingham: Open University Press.

Liberal Democrats (1994) *Reclaiming the City*, Policy Paper 2, March, London: Liberal Democrats.

Local Government Chronicle (1994) Urban regeneration bids face the axe, 9 September.

Local Government Management Board (1993) *Fitness for Purpose: Shaping New Patterns of Organisation and Management*, Luton: Local Government Management Board.

Loftman, P. and Nevin, B. (1994) Prestige project developments: economic renaissance or economic myth? A case study of Birmingham, *Local Economy*, Vol. 8, no. 4, pp. 307–25.

London Boroughs Association (1992) *Europe and the Urban Environment*, London: LBA.

Long, N. (1993) Territoriality and citizenship, *American Review of Public Administration*, Vol. 23, no. 1, pp. 19–27.

Lovering, J. (1991) Regulation/Urban Labour Markets: Towards a New Mode of Regulation? Paper to Urban Change and Conflict conference, Lancaster, September.

Lowndes, V. and Stoker, G. (1992) An evaluation of neighbourhood decentralisation; part 1 customer and citizen perspectives, *Policy and Politics*, Vol. 20, no. 1, pp. 47–61.

Mabbott, J. (1993) City Challenge – faith, hope and charities, *Town and Country Planning*, Vol. 62, no. 6, pp. 137–8.

Mackintosh, M. (1992) Partnerships: issues of policy and negotiation, *Local Economy*, Vol. 7, no. 3, pp. 210–24.

Malpass, P. (1994) Policy-making and local governance: how Bristol failed to secure city challenge funding (twice), *Policy and Politics* (forthcoming).

Marris, P. (1987) *Meaning and Action, Community Planning and Conceptions of Change*, London: Routledge and Kegan Paul.

Marsden, T., Murdoch, J., Lowe, P., Munton, R. and Flynn, A. (1993) *Constructing the Countryside*, London: UCL Press.

Marshall, T. (1994) *Survey of Cardiff Bay Businesses*, Working Paper 139, School of Planning, Oxford Brookes University, Gipsy Lane, Headington, Oxford.

Martin, S. J. (1989) New jobs in the inner city: an evaluation of the employment effects of projects assisted under the Urban Development Grant programme, *Urban Studies*, Vol. 26, pp. 627–38.

Martin, S. J. (1990) City Grants, Urban Development Grants and Urban Regeneration Grants, in M. Campbell (ed.) *Local Economic Policy*, London: Castells, pp. 44–64.

Martin, S. J. (1993) 'Economic partnership in the community: City Challenge and the Capital Partnership, Luton: Local Government Management Board.

Martin, S. J. (1994) The local authority and economic regeneration in the mid-1990s: the community, co-ordination and partnership, Luton: Local Government Management Board.

Martin, S. J. and Pearce, G. R. A. (1993) Derelict land: the case for a strategic approach, *Environmental Planning and Management*, Vol. 36, no. 2, pp. 217–29.

Massam, B. H. (1993) *The Right Place: Shared Responsibility and the Location of Public Facilities*, Harlow: Longman.

Mayer, M. (1992) The shifting local political system in European cities, in M. Dunford and G. Kafkalis (eds.) *Cities and Regions in the New Europe*, London: Belhaven Press.

Mayer, M. (1993) Urban governance in the post-Fordist city. Paper to seminar, Challenges in Urban Management, Newcastle, April.

Mayhew, P. and Maung, A. (1992) Surveying crime: findings from the 1992 British Crime Survey, *Research Finding No. 2*, Research and Statistics Department, London: Home Office.

McArthur, A. (1993), Community business and urban regeneration, *Urban Studies*, Vol. 30, no. 4–5, pp. 849–73.

McArthur, A., McGregor, A. and Stewart, R. (1993) Credit unions and low income communities, *Urban Studies*, Vol. 30, no. 2, pp. 399–416.

McConaghy, D. (1971) Inner area agencies, *Official Architecture and Planning*, May, pp. 353–6.

MC Economics Limited (1991) *The Wester Hailes Partnership: Baseline Report*, Edinburgh: MC Economics Limited.

McGregor, A., Clapham, D., Donnison, D., Gemmell, B., Goodlad, R., Kintrae, K. and McArthur, A. (1992) *Community Participation in Areas of Urban Regeneration*, Research Report No. 23, Edinburgh: Scottish Homes.

McKintosh, M. (1992) Partnership: issues of policy and negotiation, *Local Economy*, Vol. 7, no. 3, pp. 210–24.

Means, R. and Smith, R. (1988) Implementing a pluralistic approach to evaluation in health education, *Policy and Politics*, Vol. 16, no. 1, pp. 17–28.

Mega, V. (1993) Innovations for the improvement of the urban environment: a European overview, *European Planning Studies*, Vol. 1, no. 4, pp. 527–41.

Meikle, J. (1993) Lecturers indicted, *Guardian Education*, 6 July, p. 6.

Méjean (1992) Des politiques publiques contractuelles, *Informations sociales*, Vol. 19, pp. 36–43.

Meyer, P. B. (1993) Defining and measuring urban policy impacts: a comparative assessment. Paper to the Annual Meeting of the Urban Affairs Association, Indianapolis, April.

Meyer, P. (1994) Economic 'Development' and Environmental Threats: Institutional Factors Shaping Socio-Economic Impacts. Paper presented at session on 'Risk, Public Safety and the Environment', at the Sixth Annual International Conference on Socio-Economics, Paris, 15–17 July.

Mingione, E. (1991) *Fragmented Societies*, Oxford: Blackwell.

Ministère de l'Intérieur et de l'Aménagement du Territoire (1993) *Les Sociétés d'Economie Mixte Locales*, La Documentation Française, Paris.

Mishan, E. J. (1976) *Cost-Benefit Analysis*, New York, NY: Praeger.

Moore, R. (1993) Contradictions in British urban policy, *Town Planning Review*, Vol. 63, no. 3, pp. 233–4.

Moreau, J. (1989) *Administration Régionale, Départemental et Municipale* (8th edn) Mémentos Dalloz, Paris.

Morgan, D. H. J. (1972) The British Association Scandal: the effect of publicity on a sociological investigation, *Sociological Review*, Vol. 20, no. 2, pp. 185–206.

Morgan, K. (1985) Regional regeneration in Britain: the 'territorial imperative' and the Conservative state', *Political Studies*, Vol. 33, no. 4, pp. 560–77.

Morgan, K. (1992) *Resurrecting the Valleys: Can Existing Policies Do The Job?* BBC Wales, Cardiff.

Morgan, K. (1994a) Beyond the quangos: redressing the democratic deficit, *The Welsh Agenda*, Vol. 1, no. 1.

Morgan, K (1994b) *The Fallible Servant: Making Sense of the Welsh Development Agency*. Papers in Planning Research no. 151, City and Regional Planning, University of Wales College of Cardiff.

Morgan, K. and Price, A. (1992) *Rebuilding our Communities: A New Agenda for the Valleys*, Friedrich Ebert Foundation, London.

Morgan, K. and Roberts, E. (1993) *The Democratic Deficit: A Guide to Quangoland*, Papers in Planning Research no. 144, Department of City and Regional Planning, UWCC.

Morris, L. (1993) Is there a British underclass? *International Journal of Urban and Regional Research*, Vol. 17, no. 3, pp. 404–12.

Myerson, G. and Rydin Y. (1991) Language and argument in people-sensitive planning, *Planning Practice and Research*, Vol. 6, no. 1, pp. 31–3.

Nabarro, R. and Key, T. (1992) Current trends in commercial property investment and development: an overview, in P. Healey, S. Davoudi, M. O'Toole, S. Tavsanoglu and D. Usher (eds.) *Rebuilding the City*, E. and F.N. Spon, London.

National Audit Office (1990) *Regenerating the Inner Cities*, HC 169, London: HMSO.

National Audit Office (1993) *The Achievements of the Second and Third Generation Urban Development Corporation*, HC 898, London: HMSO.

NCC (1994) *Saving for Credit*, London, National Consumer Council.

National Council for Voluntary Organisations (1993) *Community Involvement in City Challenge*, London: NCVO.

Nevin, B. and Shiner, P. (1994) Behind the chimera of urban funding, *Local Work*, Vol. 52, June.

Oatley, N. (1989) Evaluation and urban development corporations, *Planning Practice and Research*, Vol. 4, no. 3, pp. 6–12.

Oatley, N. (1995) Competitive Bidding and the Regeneration Game, *Town Planning Review*, January.

OECD (1993) *Globalisation and Local and Regional Competitiveness*, OECD, Paris.

Offe, C. (1977) The Theory of the Capitalist State and the problems of policy formation, in L. N. Linberg and A. Alford (eds.) *Stress and Contradiction in Modern Capitalism*, Lexington, Mass.

Osborne, D. and Gaebler, T. (1993) *Reinventing Government*, New York: Plume Books.

Osmond, J. (ed.) (1994) *A Parliament For Wales*, Llandysol, Gower Press.

Pacione, M. (1990) What about people? A critical review of urban policy in the United Kingdom, *Geography*, Vol. 75, pp. 193–201.

Parkinson, M. (1989) The Thatcher government's urban policy 1979–1989: a review, *Town Planning Review*, Vol. 60, no. 4, pp. 421–40.

Parkinson, M. (1990) Political responses to urban restructuring: the British experience under Thatcherism, in J. Logan and T. Swanstrom (eds.) *Beyond the City Limits*, Philadelphia: Temple University Press.

Parkinson, M. (1992) City Challenge. Paper to TCPA seminar, Urban Regeneration, Newcastle, 22 November.

Parkinson, M. (1993) A new strategy for Britain's cities? *Policy Studies*, Vol. 14, no. 2, pp. 5–13.

Pearce, G. R. A. (1988) City Grants: lessons from the Urban Development Grant programme, *Journal of the Royal Town Planning Institute*, Vol. 74, no. 4, pp. 15–19.

Pearce, G. and Martin, S. (1994) *The measurement of additionality: grasping the slippery eel*. Paper presented to the ESRC Urban Policy Evaluation seminar, University of Wales, Cardiff, 7–8 September.

Pearce, J. (1993) *At the Heart of the Community Economy*, London: Calouste Gulbenkian Foundation.

Peck, J. and Tickell, A. (1992) Local modes of social regulation? Regulation theory, Thatcherism and uneven development, *Geoforum*, Vol. 23, no. 3, pp. 347-63.

Peck, J. and Tickell, A. (1993) Business goes local: dissecting the 'Business Agenda' in post-democratic Manchester. Paper presented at the Ninth Urban Change and Conflict Conference, University of Sheffield, September.

Perry, J. (1991) Swopping new myths for old, *Housing*, February.

Plaid Cymru (1993) *A New Deal for the Valleys*, Cardiff: Plaid Cymru.

Planning 25 September 1992 'Challenge Revolution in Inner City Attitudes' p. 8.

Platt, J. (1987) The contribution of social science, in M. Loney *et al.* (eds.) *The State and the Market: Politics and Welfare in Contemporary Britain*, London: Sage, pp. 236–47.

Policy Studies Institute (1994) *Urban Trends II*, London: PSI.

Price Waterhouse (1993) *Evaluation of Urban Development Grant, Urban Regeneration Grant and City Grant*, London: HMSO.

Punter, J. (1993) Development interests and the attack on planning controls: 'planning difficulties' in Bristol 1985–1990, *Environment and Planning A*, Vol. 25, pp. 521–38.

Reade, E. (1982) Section 52 and corporatism in planning, *Journal of Planning and Environment Law*, pp. 8–16.

Redwood, J. (1993) Speech to the Society of Chief Personnel Officers, Scarborough, 25 March.

Rees, G. (1989) The Valleys Programme: some problems of evaluation. Paper to the Wales TUC Conference on The Valleys Programme: One Year On, Polytechnic of Wales, 23 June.

Roberts, P. (1990) Strategic planning and management: new tasks in the 1990s, *Journal of the Royal Town Planning Institute*, Vol. 76, no. 49, pp. 25–30.

Roberts, T. (1991) *The Valleys Initiative: Three Years On*, Council of Welsh Districts, Cardiff.

Robinson, F. and Shaw, K. (1991) Urban regeneration and community involvement, *Local Economy*, Vol. 6, no. 1, pp. 61–73.

Robinson, I. *et al.* (1993) *More Than Bricks and Mortar*, Durham: University of Durham.

Robson, B.T., Bradford, M. G., Deas, I., Hall, E., Harrison, E., Parkinson, M., Evans, R., Garside, P. and Harding, A. (1994) *Assessing the Impact of Urban Policy*, Department of the Environment, Inner Cities Research Programme, London: HMSO.

Rogerson, R., Findlay, A., Morris, A. and Coombes, M. G. (1989) Indicators of quality of life: some methodological issues, *Environment and Planning A*, Vol. 21, pp. 1655-66.

Rose, R. (1993) *Lesson-Drawing in Public Policy*, New Jersey: Chatham House Publishers.

Ross, B. H., Levine, M. A. and Stedman, M. S. (1991) *Urban Politics: Power in Metropolitan America*, (4th edn) Itasca, Illinois: F. E. Peacock.

Rossi, P. and Freeman, H. (1993) *Evaluation: A Systematic Approach 5*, London: Sage.

Rydin, Y. (1993) *The British Planning System*, London: Macmillan.

Sackman, H. (1974) *Delphi Assessment: Expert Opinion, Forecasting and Group Procedure*, Santa Monica, CA: Rand Corporation.

Sayer, A. (1992) *Method in Social Science* (2nd edn) London: Routledge.

Scarman, L. G (1981) *The Scarman Report. The Brixton disorders 10–12 April 1981*, London: HMSO.

Scottish Office (1988) *New Life for Urban Scotland*, Edinburgh: Scottish Office.

Scottish Office Industry Department (1993) *Strathclyde IDO Interim Evaluation*, a report by Pieda plc. Planning, Economic and Development Consultants, Edinburgh.

Sen, A. K. (1982) *Choice, Welfare and Measurement*, Oxford: Basil Blackwell.

Shelter Neighbourhood Action Project (1972) *Another Chance for Cities*, London: Shelter.

Smith, G. and Cantley, C. (1985) *Assessing Health Care: a Study in Organisational Evaluation*, Milton Keynes: Open University Press.

Smith, S. J. (1989) *The Politics of 'Race' and Residence: Citizenship, Segregation and White Supremacy in Britain*, Cambridge: Polity Press.

Solesbury, W. (1987) Urban policy in the 1980s: the issues and arguments, *Journal of the Royal Town Planning Institute*, Vol. 73, no. 6, pp. 18–22.

Solesbury, W. (1993) Reframing urban policy, *Policy and Politics*, Vol. 21, no. 1, pp. 31–8.

South Glamorgan County Council (1986) *The Regeneration of South Cardiff: Proposals for an Urban Development Corporation*, Cardiff: South Glamorgan County Council.

Spicer, J. S. and Grigg, A. R. (1980) *An Analysis of the Usage of a Property Information System*, London: HMSO.

Stewart, J. (1989) The changing organisation and management of local authorities, in J. Stewart and G. Stoker (eds.): *The Future of Local Government*, London: Macmillan.

Stewart, J. (1992) *The Rebuilding of Public Accountability*, London: European Policy Forum.

Stewart, M. (1987) Ten years of inner cities policy, *Town Planning Review*, Vol. 58, no. 2, pp. 129–45.

Stewart, M. (1988) The finding of Wigan Pier? A review article on the ESRC Inner Cities Research Programme, *Policy and Politics*, Vol. 16, no. 2, pp. 123–32.

Stewart, M. (1990) Urban Policy in Success City: the application of inner area policies to Bristol, in *Urban Policy in Thatcher's England*, School for Advanced Urban Studies, University of Bristol, Working Paper 90.

Stewart, M. (1993) Value for money in urban public expenditure. Paper presented at the ESRC Urban Policy Evaluation Seminar at the University of Wales College, Cardiff, 27–28 September. Available from the Short Course Secretary, Department of City and Regional Planning, University of Wales College of Cardiff, PO Box 906, Cardiff CF1 3YN.

Stewart, M. (1994a) Between Whitehall and Town Hall: the realignment of regeneration policy in England, *Policy and Politics*, Vol. 22, no. 2, pp. 133–46.

Stewart, M. (1994b) Value for money in urban public expenditure, *Public Money and Management*, Vol. 14, no. 4,

Stewart, M. and Leach, S. (1992) *The Future Role and Function of Local Government*, York: Joseph Rowntree Foundation.

Stoker, G. (1988) *The Politics of Local Government*, London: Macmillan.

Stoker, G. (1990) Regulation theory, local government and the transition from Fordism, in D. King and J. Pierre (eds.) *Challenges to Local Government*, London, Sage, pp. 242–64.

Stoker, G. and Wolman, H. (1992) Drawing lessons from US experience: an elected mayor for British local government, *Public Administration*, Vol. 70, Summer, pp. 241–67.

Stoker, G. and Young, S. (1993) *Cities in the 1990s*, London: Longman.

Storey, D. (1990) Evaluation of policies and measures to create local employment, *Urban Studies*, Vol. 27, pp. 669–84.

Strathclyde Regional Council (1994) *Strathclyde Economic Trends*, no. 42, April, Glasgow: Strathclyde Regional Council.

Sugden, R. and Williams, A. (1978) *The Principles of Practical Cost-Benefit Analysis*, New York, NY: Oxford University Press.

Svara, J. H. (1990) *Official Leadership in the City*, Oxford University Press.

Thain, C. and Wright, M. (1990) Coping with difficulty: the Treasury and public expenditure 1976–79, *Policy and Politics*, Vol. 18, no. 1, pp. 1–16.

Thake, S. and Staubach, R. (1993) *Investing in People: Rescuing Communities from the Margin*, York: Joseph Rowntree Foundation.

The Royal Town Planning Institute (1993) *The Impact of the European Community on Land Use Planning in the United Kingdom*, Centre for European Property Research, University of Reading.

The Royal Town Planning Institute (1994) *The impact of the European Community on Land Use Planning in the United Kingdom*, Centre for European Property Research, University of Reading.

Thomas, H. (1989) Evaluating CBDC's effectiveness, *The Planner*, Vol. 75, no. 1, pp. 26-7.

Thomas, H. (1992) Mutiny on the boundaries, *Surveyor*, Vol. 177, no. 5185, pp. 8–9.

Thomas, H. and Imrie, R. (1989) Urban redevelopment, compulsory purchase and the regeneration of local economies, *Planning Practice and Research*, Vol. 4, no. 3, pp. 18–27.

Thomas, H. and Imrie, R. (1993) Cardiff Bay and the project of modernisation, in R. Imrie and H. Thomas (eds.) *British Urban Policy and the Urban Development Corporations*, London: Paul Chapman, pp. 74–88.

Thompson, M. S. (1980) *Benefit-Cost Analysis for Programme Evaluation*, Beverly Hills, CA: Sage Publications.

Thornley, A. (1991) *Urban Planning under Thatcherism*, London: Routledge.

Toynbee, P. (1993) Report on *Today*, Radio 4, 9 June.

Trust in the Community (1992) *City Challenge Second Round Bid Document*.

Turok, I. (1989) Evaluation and understanding in local economic policy, *Urban Studies*, Vol. 26, pp. 587–606.

Turok, I. (1991) Policy evaluation as science: a critical assessment, *Applied Economics*, Vol. 23, pp. 1543–50.

Victor Hausner and Associates (1993) *The Programme for the Valleys: An Evaluation*, Cardiff: Welsh Office.

Ville de Lille (1989) *Plan Local de Développement Social de Moulins*, Lille.

Wainwright, M. (1991) Riot puts despair on agenda, *The Guardian*, 11 September.

Ward, M. (1981) Progress towards indicator systems: an overview, in C. L. Taylor (ed.) *Indicator Systems for Political, Economic and Social Analysis*, Cambridge,

Mass: Gunn and Hain.

Ward, R. with Randall, R. and Wilson, P. (1989) *Evaluation of the Ethnic Minority Business Initiative: Final Report, Working Draft*, London: Home Office.

Warren, R., Rosentraub, M. S. and Weschler, L. F. (1992) Building urban governance: an agenda for the 1990s, *Journal of Urban Affairs*, Vol. 14, no. 3/4, pp. 399–422.

Weir, S. (1994) *Ego Trip: Extra-Governmental Organisations in the United Kingdom and their Accountability*, London: Charter 88 Trust.

Welsh Development Agency (1992) *Urban Development Wales*, Cardiff.

Welsh Office (1988) *The Valleys: A Programme for the People*, Cardiff.

Welsh Office (1993a) *Programme for the Valleys: Building on Success*, Cardiff.

Welsh Office (1993b) *Programme for the Valleys: a Statistical Profile*, Cardiff.

Wester Hailes Partnership (1989) *Realising the Potential: the Partnership Strategy for Wester Hailes*, Edinburgh: Scottish Office.

Wester Hailes Partnership (1993) *Wester Hailes Census Profile*, Edinburgh: Wester Hailes Partnership.

Wester Hailes Representative Council (1989) *The Pitlochry Affirmation*, Edinburgh: Wester Hailes Representative Council.

Whitney, D. and Haughton, G. (1990) Structures for development, partnerships in the 1990s, *The Planner*, June, pp. 15–19.

Wilkinson, S. (1992) Towards a new city? A case study of image-improvement initiatives in Newcastle-upon-Tyne, in P. Healey *et al.* (eds.) *Rebuilding the City*, London: E. and F. N. Spon, pp. 174–211.

Williams, B (1973) A critique of utilitarianism, in J. J. C. Smart and B. Williams *Utilitarianism: For and Against*, pp. 75–150, London: CUP.

Williams, F. (1994) Social relations, welfare, and the post-Fordism debate, in R. Burrows (ed.) *Towards a Post-Fordist Welfare State*, London: Routledge, pp. 49–73.

Willmott, P. (1989) *Community Initiatives – Patterns and Prospects*, London: Policy Studies Institute.

Willmott, P. (ed.) (1994) *Urban Trends 2: a Report on Britain's Deprived Urban Areas*, London: PSI.

Willmott, P. and Hutchinson, R. (eds.) (1992) *Urban Trends 1*, London: PSI.

Wolman, H. (1993) Cross-national comparisons of urban economic programmes: is policy transfer possible? in D. Fasenfest (ed.) *Community Economic Development Policy Formation in the US and the UK*, Macmillan, London.

Wolman, H. and Goldsmith, M. (1992) *Urban Politics and Policy*, Oxford: Blackwell.

Wong, C. (1993) Towards better practice in planning information collection, *Town and Country Planning*, Vol. 62, no. 10, pp. 279–81.

Wintow, P. (1994) Treasury seeks to end industrial aid, *The Guardian*, 1 August.

Young, M. (1991) *An Inside Job*, Oxford: Clarendon Press.

Young, S. C. (1993) Assessing the Challenge's local benefits, *Town and Country Planning*, January/February, pp. 3–5.

Yount, K. and Meyer, P. B. (1994) Bankers, developers and new investment in brownfield sites: environmental concerns and the social psychology of Risk', *Economic Development Quarterly*, November.

INDEX